ORGANIZATIONAL BEHAVIOR AND VIRTUAL WORK

Concepts and Analytical Approaches

ORGANIZATIONAL BEHAVIOR AND VIRTUAL WORK

Concepts and Analytical Approaches

Arvind K. Birdie, PhD
Madhu Jain, PhD

Apple Academic Press Inc. | Apple Academic Press Inc.
3333 Mistwell Crescent | 9 Spinnaker Way
Oakville, ON L6L 0A2 | Waretown, NJ 08758
Canada | USA

©2016 by Apple Academic Press, Inc.

First issued in paperback 2021

Exclusive worldwide distribution by CRC Press, a member of Taylor & Francis Group
No claim to original U.S. Government works

ISBN 13: 978-1-77463-589-6 (pbk)
ISBN 13: 978-1-77188-283-5 (hbk)

Library and Archives Canada Cataloguing in Publication

Birdie, Arvind K., author
Organizational behavior and virtual work : concepts and analytical approaches /
Arvind K. Birdie, PhD, Madhu Jain, PhD.

Includes bibliographical references and index.
Issued in print and electronic formats.
ISBN 978-1-77188-283-5 (hardcover).--ISBN 978-1-77188-284-2 (pdf)

1. Telecommuting. 2. Work environment. 3. Work-life balance. 4. Organizational change. 5. Organizational behavior. I. Jain, Madhu (Professor), author II. Title.

HD2336.3.B57 2016   658.3'123   C2016-902440-7   C2016-902441-5

Library of Congress Cataloging-in-Publication Data

Names: Birdie, Arvind K., author.
Title: Organizational behavior and virtual work : concepts and analytical approaches / Arvind K. Birdie, PhD, Madhu Jain, PhD.
Description: Toronto ; New Jersey : Apple Academic Press, 2015. | Includes bibliographical references and index.
Identifiers: LCCN 2016016121 (print) | LCCN 2016017538 (ebook) | ISBN 9781771882835 (hardcover : alk. paper) | ISBN 9781771882842 ()
Subjects: LCSH: Organizational behavior. | Virtual work teams. | Virtual reality in management.
Classification: LCC HD58.7 .B558 2015 (print) | LCC HD58.7 (ebook) | DDC 302.3/5--dc23
LC record available at https://lccn.loc.gov/2016016121

Apple Academic Press also publishes its books in a variety of electronic formats. Some content that ap-pears in print may not be available in electronic format. For information about Apple Academic Press products, visit our website at **www.appleacademicpress.com** and the CRC Press website at **www.crc-press.com**

ABOUT THE AUTHORS

Arvind K. Birdie, PhD

Arvind K. Birdie, PhD, is Associate Professor and Acting Principal at IIMT School of Management (Oxford Brooks University), Gurgaon, India. Dr. Birdie has been consistently recognized for her teaching abilities. Her strength lies in teaching various interdisciplinary subjects with equal ease. Besides academic teaching and training, she has organized management development programs for corporate and academicians. She is a regular presenter at various international and national conferences and has published several papers in refereed journals. Her areas of interest include leadership, work-life balance, virtual work, and positive psychology. She has been taking postgraduate and graduate courses of management for more than 13 years.

Madhu Jain, PhD

Madhu Jain, PhD, is presently Associate Professor in the Department of Psychology at the University of Rajasthan, Jaipur, India. She is also Director of the Department of Life Long Learning, also at the University of Rajasthan. She has been teaching and guiding postgraduate, graduate, and PhD students for over two decades and has supervised several research projects at the advanced level as well. Dr. Jain has published nearly forty papers in scholarly national and international journals and also published four books in the area of psychology. Dr. Jain is associated with many national and international associations. As a student, she was a Gold Medalist and Merit Holder at the graduate and post graduate level.

DEDICATED

*With love to my source of inspiration and courage - Parents,
Gagan and Shaurya !*

—AKB

CONTENTS

LIST OF ABBREVIATIONS

ANOVA	analysis of variance
APEC	Asia-Pacific Economic Co-operation Forum
BSB	boundary spanning behavior
CAOC	cynicism about organizational change
CFA	confirmatory factor analysis
CI	continuous improvement
CMCS	computer-mediated communication system
EU	European Union
GOCB	group organizational citizenship behavior
GDSS	group decision support system
GDT	geographically dispersed team
HIWPs	high-involvement work practices
ICT	information and communications technologies
KRI	Kenexa Research Institute
LBDQ	leader behavior description questionnaire
LES	leadership effectiveness scale
LMX	leader–member exchange
LSE	leadership self-efficacy
LSI	leader's ability to support innovation
MBO	management by objectives
MLMs	multilevel marketing organizations
NAFTA	North American Free Trade Agreement
N-Affil	need for affiliation
NCR	National Capital Region
nPow	need of power
OCB	organizational citizenship behaviors
OTI	Organizational Trust Inventory
PD	power distance
PDI	power distance index
PDM	participative decision-making
POS	perceived organizational support
RTCs	residential treatment centers

SBA   Small Business Administration
SEM   structural equation modeling
SDWT   self-directed work teams
SMT   self-managed work team
TLB   transformational leader behaviors
TPS   Toyota Production System
UNDP   United Nations Development Program
VCR   virtual collaborative relationship
WABA   within and betweenanalysis

PREFACE

Organizational Behavior and Virtual Work: Concepts and Analytical Approaches is the blend of theory and application in the real world of virtual workers. It reflects the most recent research and developments in the new work–employment relationship of virtual work. It takes an in-depth, critical look into the key factors affecting virtual work environments with practical input of suggestions and recommendations.

• WHY STUDY VIRTUAL WORK?

With globalization and technological advancement and emphasis on service quality today, organizations have changed a lot in recent years. For balancing work life in the present era of working 24×7, the new employment relationship of "virtual work" has emerged and was fast picked up by the employees in the field of organizational behavior. Keeping in perspective the effect of virtual work environment's where people work in teams and face-to-face meetings are missing, this book explores the dynamics of changing organization structures, theories of leadership, trust, and how dimensions of self-efficacy work in this new work relationship. This book caters to the needs of researchers, policymakers, managers, and students to understand better the future work relationship known as the virtual work concept and how the organizational behavior dimensions—importantly, leadership, trust, self-efficacy, and organization climate—are being affected by this new employment relationship as very little research is available specifically from a South Asian perspective.

• OBJECTIVE OF THE BOOK

The main objective of this text is to create and build a reader-friendly and comprehensive knowledge base that gives an engaging introduction of virtual work and organizational behavior's important aspects, which are in a transitional stage. This book covers both theoretical and empirical bases of virtual work aspects, underlining application-based suggestions for managers, so it caters to students, researchers, policymakers, as well as managers.

• THE STRUCTURE OF THE BOOK

This book elucidates behavior at two levels, starting with providing a strong understanding of theoretical concepts and then providing research on the correlation between different variables of organizational behavior in virtual workers.

PART I
INTRODUCTION

CHAPTER 1

ORGANIZATIONS

CONTENTS

1.1   THE ESSENCE

In order to understand forces around us, we need to see with the nature of "organizations" as social phenomenon. Simply stated, "an Organization is a collection of people working together to achieve a common purpose." It describes everything from clubs, voluntary organizations, and religious bodies to entities, such as large and small businesses, unions, hospitals, schools, and government agencies.

Organizations have always existed in one form or another. When human beings began to do collective activity as a means of improving their chances of survival and quality of life, the basis of the social organization was formed. This form of activity is still evident today in the many family businesses (Bork, 1986). There is evidence of early forms of organizations not based upon the family unit being used by the Sumerian people who settled around the river Euphrates approximately 3500 years BC (McKelvey, 1982). Of course, many organizations during that period continued to be family run. The leadership of the state was based upon dynastic family groupings.

Organizations can also be categorized into those that operate for profit and not for profit institutions. An organization could be described as a collection of individuals who are organized into groups and subgroups and "interact" with each other in an "interdependent relationship." The individuals work toward "common goals," which are not always clear, and the way they relate is determined by the structure of the organization (Duncan, 1981).

Perrow (1970) mentioned five organizational goals:

> Societal goals: These goals refer to the production of goods or provision of services, the maintenance of cultural values in the society.
>
> Output goals: These goals are designed to satisfy customer or client needs for goods or services as this could be viewed as the primary task of the organization.
>
> System goals: These goals are used to promote the interests of the organization independent of its output. For example, these goals could be in the form of an intention to augment the organization growth or as a market leader.

Product characteristic goals: These goals could refer to the production of a distinctive product with style and they could be derived from either output or system goals.

Derived goals: These goals apply when top management uses the organization's power and resources to advance social and cultural causes etc.

March and Simon (1958) drew a distinction between three types of goals:

Official goals: These goals are stated in the charter of the organization or established by the founder of the organization.

Operative goals: These goals reflect the real intentions of the leaders or dominant coalitions within the organizations and are concealed from the public.

Operational goals: These goals could be expressed in quantitative terms. Whereas growth of the organization may be an operative goal, it becomes an operational goal when it is expressed in percentage terms (e.g., 7% growth rate).

1.2 TRADITIONAL ORGANIZATIONS

Organizations and their management can be traced from the history itself. A large number of archaeological evidence and big monuments like TajMahal, The Great Pyramids, and cultures and civilizations like Indus valley are few examples of prescientific era organizations. Since ancient times, traditional organizations were more of authoritarian, which stands in silent tribute to the talents of workers who were in miserable conditions and according to historical and archaeological evidences, were even harshly treated after taking work from them. Monument like Taj Mahal is a big example where King Shahjahan had ordered for cutting the hands of numerous workers who made Taj Mahal beautifully by working hard in difficult working conditions. As the pyramid shape suggests, power was concentrated primarily among the handful of individuals at the top.

As time passed, natural resources exploration and advancement in tools led to Industrial Revolution, which bought big change through migration and change in quality of life in people. People started taking education and moved to cities from villages to work in factories. This phase came as the starting of Classical Management era. Now there was a level of workers, economy, tools, techniques, and roles in comparison to earlier period of AD 1000–1880. In the Classical era, two major thrusts emerged as administrative theory approach traced from Henri Fayol who divided managers' roles into five functions: planning, organizing, command, coordination, and control. His 14 principles of management emphasized division of labor, authority, discipline, and a strictly enforced chain of command. Secondly in 1911, Frederick Winslow Taylor in his book, *Principles of Scientific Management*, introduced the principles for designing and managing mass-production facilities such as Ford's automobile factory in Michigan and Carnegie's steel works in Pittsburgh. He saw workers basically as lazy beings, motivated primarily by money. The emergence of a factory system of production during the early stages of industrial capitalist development in Europe and the United States presaged the beginning of organizational conflict. A traditional way of life and labor was disrupted. This provoked intense resistance, opposition, and conflict over the emerging organization of factory production (see Pollard, 1965; Thompson, 1963; Bendix, 1956; Montgomery, 1979; Gutman, 1975). As increasingly larger portions of the population are forced into the labor market, where they must sell their labor power for a wage, the proletariat or working class is created. A large mass of workers are now organizationally constrained within a hierarchical factory system. A large part of the evolution of organization theory and management strategy can be chronicled as a history of trial and error in developing methods and techniques for this control and extraction.

At the time, however, the monumental challenge of coordinating and controlling large numbers of workers within a single factory had never been confronted on such a scale. During this period, one of the most significant sources of conflict, according to Bendix (1956), was "traditionalism"– the ideological way of life among labor prescribing precapitalist customs, norms, routines, and work habits. This stood as the major obstacle to the enforcement of the "new discipline" within the factory. In the United States the heterogeneity of the labor factor, fueled by the constant flow of immigrants, resulted in a variety of cultural habits that did not fit smoothly into the emerging industrial machine (Gutman, 1975; Montgomery, 1979).

Thus, the factory organization was characterized by an array of competing forces–traditional work habits, an emerging production system, managerial strategies to break traditions and impose discipline, and the reaction and resistance of labor (Jeffy, 2008).

During the 1930s, a unique combination of factors developed the emergence of new era called Behavioral era. The legalization of union management collective bargaining in the United States in 1935 forced management to opt for new ways of handling employees and at the same time, behavioral scientists started emphasizing on "human factor." Western Electric's Hawthorne plant study was a prime stimulus for this human relations management enforcement. A general perspective of the Behavioral era shows starting of organization's focus on humane work environment and informal work groups.

1.3 CULTURAL PERSPECTIVES

Culture has been the subject of investigation in social anthropology, where researchers have sought to understand the shared meanings and values held by groups in society that give significance to their actions (McKenna, 2000). Today, many organizations are sensitive to the fact that cultural differences exist and how it plays dynamic role in communication, climate, and behavior in organizations. SouthAsian countries like Japan where people derive great satisfaction and happiness from relationships and working in groups, American culture puts more emphasis on individual goals and independence. The distinction between individualism and collectivism makes sense when we realize that people across all cultures define themselves in terms of both their uniqueness, that is, personal identity and their relationship to others (social identity). Some cultures clearly reflect more than the others, but both have a place in a person's values and self-concept (Mc Shane, Glinow, & Sharma, 2011).

In Western cultures, predictors of happiness, include elements that support personal independence, a sense of personal agency, and self-expression. In Eastern cultures, predictors of happiness focus on an interdependent self that is inseparable from significant others. Compared to people in individualistic cultures, people in collectivistic cultures are more likely to base their judgments of life satisfaction on how significant others appraise their life than on the balance of inner emotions experienced as pleasant versus unpleasant.

Geert Hofstede carried out one of the most comprehensive studies on how values in the workplace are influenced by culture. Hofstede is a renowned Dutch organizational sociologist who conducted detailed interviews with thousands of IBM employees in 53 countries from 1978–1983 and has continued to develop his research since. In Japanese organizations, supervisors and employees have a largely egalitarian relationship where consensus on both parts is required for making decisions. Rather than being a source of authority, top management is seen as a facilitator/ consensus builder and has the responsibility of maintaining harmony so that employees can work together. Top management takes cues from middle management, who base policies on the information forwarded by subordinates.

The Toyota Production System (TPS) (despite recent issues) is probably the most famous and successful example of a continuous improvement (CI) culture (CI is the ongoing effort to improve products, services, and processes). Toyota's founder, Kiichiro Toyoda, is a firm believer that "each person fulfilling his or her duties to the utmost can generate great power when gathered together, and a chain of such power can generate a ring of power."

"Hourensou" (hou = report, ren = inform, and sou = feedback) means to report out to other's frequently and keep those necessary informed of your work, while remaining open to feedback and direction from peers. "Genchigenbutsu" means "getting your hands dirty, to identify or solve immediate problems and leaders are not exempt from this."

In North America, a person can bounce from career to career, from being a chartered accountant to a band roadie, but in Japan, similar to other Asian cultures, employees are expected to stay with one company for their entire working careers. In Japanese organizations, supervisors and employees have a largely egalitarian relationship where consensus on both parts is required for making decisions. Rather than being a source of authority, top management is seen as a facilitator/consensus builder and has the responsibility of maintaining harmony so that employees can work together. Top management takes cues from middle management, who base policies on the information forwarded by subordinates.

The decision-making process within Chinese firms is based on respect, evasiveness, hierarchy, and discipline. Open conflict is avoided at all costs, even if upper management is clearly making a wrong decision. Unlike our previous Asian counterparts, individualism (IDV) is significantly higher in Germany, with a much higher valuation on people's time and freedom. The masculinity index (MAS) also ranks quite high, not because men are particularly dominating the workforce but rather, Germans as a people, value competitiveness, assertiveness, ambition, and the accumulation of wealth and material possessions. Swedish companies tend to have a flat and team-oriented structure with few management levels, reflected in the moderate power distance index (PDI)). When looking at Sweden's performance appraisal, like other Scandinavian countries, they rank very high for IDV, indicating its population is for the most part self-reliant, enjoys challenges, and respects and values privacy. Sweden is also known to be a very open, nurturing, and egalitarian society, and this is clearly reflected in the low MAS ranking. Again, this is a manifestation of a culture that embraces and values quality of life rather than quantity of possessions and achievements. Out of all the countries analyzed in Hofstede's research, only seven have IDV as their highest dimension, including the United States (91), Australia (90), the United Kingdom (89), the Netherlands and Canada (80), and Italy (76). While the globalization of businesses may carry us closer to universal standards for work, in comparing these distinctive management styles, there is a clear and present need for organizations to acclimatize to diversity. By incorporating cultural aspects into everyday business, and by developing strong and effective international management teams, companies will have the competitive advantage to become successful global players (Gallant, 2010).

1.4   ELEMENTS OF ORGANIZATIONS

1.4.1   ORGANIZATION DESIGN

Organization design is the process by which managers create a specific type of organizational structure and culture so that a company can operate in the most efficient and effective way (Child, 1977).

1.4.2   ORGANIZATIONAL STRUCTURE

Organizational structure is the formal system of task and job reporting relationships that determines how employees use resources to achieve organizational goals(Jones & George 2003).

1.4.2.1   FACTORS AFFECTING ORGANIZATIONAL STRUCTURE:

- Organizational environment
- Technology
- Strategy
- Human resources

Small companies can use a variety of organizational structures. However, a small company's organization structure must be designed to effectively meet its goals and objectives, according to the Lamar University article titled "organizational structure" on its website. Types of organizational structure in management can include flat structures, as well as functional, product, and geographical-structured organizations.

1.4.2.2   FLAT ORGANIZATIONAL STRUCTURE

Many small companies use a flat organizational structure, where very few levels of management separate executives from analysts, secretaries, and lower-level employees. Flat organizations work best when a company has less than 20 employees, especially if the company employs one or two employees per department. One advantage of using a flat organizational structure for management is that decisions can be made relatively quickly. The flat organizational structure lacks the typical bureaucracy of taller organizational structures–those with many levels of management.

1.4.2.3   FUNCTIONAL ORGANIZATIONAL STRUCTURE

A functional organizational structure is centered on job functions, such as marketing, research and development, and finance. Small companies should use a functional organization when they want to arrange their organizational structure by department. For example, a small company may have a director, two managers, and two analysts in the marketing department. The director would likely report to the chief executive officer, or CEO, and both managers would report to the director. In addition, each manager may have an analyst reporting to them. A functional organizational structure works well when small companies are heavily project-focused. Directors can assign certain projects to managers, who can then divvy up tasks with their analysts. The department can then more effectively meet their project deadlines.

1.4.2.4 PRODUCT ORGANIZATIONAL STRUCTURE

A product organizational structure has managers reporting to the president or head of the company by product type. Product organizational structures are primarily used by retail companies that have stores in various cities. However, stores in each city may still need local human resources or marketing department to carry out functions locally. For example, a small department store company may have a vice president of sporting goods, housewares, and general merchandise at the corporate office. One manager may report to each vice president. However, each manager may oversee the work of one or more field marketing employees who travel and handle local marketing stores in several states. These field marketing employees may work for the sporting goods manager 1 week in League City, Texas, then do merchandising for the housewares manager another week in the Sugarland, Texas, market.

1.4.2.5 GEOGRAPHICAL ORGANIZATIONAL STRUCTURE

The small business administration (SBA) is responsible for defining small businesses in different industries. For example, in manufacturing, the SBA usually considers a company with 500 or fewer employees a small business. The point is small businesses are still large enough to use a geographical organizational structure. A geographical organizational structure is when companies decentralize the functional areas. For example, unlike the product organizational structure, there may be a local marketing, finance, accounting, and research development person based in each region. For example, a small consumer products food company may be large enough to place a marketing research manager and analyst in each of six different regions. This can be important because consumers in various areas have different tastes. Hence, a geographical structure will enable the company to better serve the local market.

The basic concepts of organization design were formulated in the early 1900s by management writers who offered numerous insights in the form of principles for managers to follow and still these principles are majorly valuable for designing effective and efficient base of organizations inspite of their limitations with changing times over the years.

1.4.3 ORGANIZATION DESIGN APPLICATIONS

Since old days, organizations evolved in one or some kind of designs, which is changing in recent years at a fast pace due to technological advances and other factors like globalization. Letus look at the various types of organization design starting from simple and conventional.

1.4.3.1 SIMPLE STRUCTURE

Major organizations start their journey from a simple structure. Like a simple shop owner with sales person directly reporting to him. As organizations grow they develop complex structures. In simple structure, authority lies centrally to one person—the owner. The simple structure is mostly seen in small business organizations. Accountability is clear and it is a flat organization usually having two or three vertical levels. As size becomes large, decision-making becomes slower and organization becomes more formal. In a simple structure organization everything depends on one person and is usually a single man show and risky.

1.4.3.2 BUREAUCRACY

Bureaucracy refers to type of structure characterized by division of labor, a well-defined authority hierarchy, high formalization, impersonal relations, employment decisions based on merit, career tracks for employees, and distinct separation of member's organization and personal lives. The strength of the bureaucratic form lies in standardization. The organization is more efficient as behavior of employee is controlled and predictable. On the downside bureaucracy leads employee alienation, the concentration of power in the hands of a few, and frustrations for the clients who must deal with rule bound behavior. It is the dominant organizational form in society and has achieved its distinction because it works best with the type of technologies and environment that most organizations have.

1.4.3.3   FUNCTIONAL STRUCTURE

Organizational structureisa structure that groups similar or related occupational specialties together. It merely expands the functional orientation to make it the dominant form for the organization. The strength of the functional structure lies in the advantages that accrue from work specialization.

1.4.3.4   DIVISIONAL STRUCTURE

This organizational structure is made up of separate business units or divisions. In this structure, each division has limited autonomy, with a division manager who has authority over his/her unit and is responsible for performance.

The element of structure acts as an external overseer to coordinate and control the various divisions and it usually provides services such as financial and legal support services.

1.5   CHALLENGES OF ORGANIZATIONAL DESIGN

To be an effective organization as it changes and grows with the changing environment, managers have to continuously evaluate the way their organization is designed; to say, how work is divided among people and departments, how it utilizes the human, financial, and physical resources.

1.5.1   DIFFERENTIATION

It is the process of allocating people and resources for organizational tasks which establishes the task and authority relationship that allows the organization to achieve its goals. In brief, it establishes and controls the division of labor or degree of specialization in the organization. In a simple organization, differentiation is low because the division of labor is low. As one person or a few people do all organizational tasks, there are few problems with coordinating who does what for whom and when. With expansion comes complexity.

1.5.2   ORGANIZATIONAL ROLES

The base of differentiation is organization role which is task related behavior required/expected by position of person in an organization. As the division of labor increases in an organization, managers specialize in some roles and for other specialized role hire people. Specialization helps executives to develop their individual abilities and knowledge, which helps in achieving organization's ultimate goal. A system of interlocking roles and their relationship becomes base of organizational structure.

1.5.3   CENTRALIZATION:

Decisions are made in the organization: either on top by senior management or down low. Most theorists concur that the term refers to the degree to which decision-making is concentrated at a single point in the organization. A high concentration implies high centralization, whereas a low concentration indicates low centralization which may be called as decentralization.

1.6   DIMENSIONS OF STRUCTURE

1.6.1   COMPLEXITY

It is the degree of differentiation that exists within an organization. Horizontal differentiation considers the degree of horizontal separation between units. Vertical differentiation refers to the depth of organization hierarchy. Spatial differentiation encompasses the degree to which the location of an organization facilities and personnel are dispersed geographically. An increase in anyone of these three factors will increase an organization complexity.

1.7   CONTEMPORARY ORGANIZATIONAL DESIGNS

1.7.1   MATRIX AND PROJECT STRUCTURES

In matrix structure, specialists from different functional departments work on projects that are led by project manager and this structure creates a dual chain of command in which employees have to report to two managers, one functional and other product. Some organizations use project structure, in which employees continuously work on projects. Project structure has no formal departments and it works as a flexible organizational design. There is no rigid organizational hierarchy to slow down decision-making process or action.

Advantages of the matrix include (1) increased ability to respond rapidly to change in the environment, (2) an effective, means for balancing the customers required for project completion and cost control with the organization's need for economic efficiency and development of technical capability for future, and (3) increased motivation by providing an environment more in line with the democratic norms preferred by scientific and professional employees. The main disadvantage of the matrix lies in the confusion it creates, its propensity to foster power struggle and the stress it places on an individual.

1.7.2   TEAM STRUCTURES

A team structure is one in which the entire organization is made up of work teams that do the organization's work. Employee empowerment is very important for this type of structure as there is no hierarchy. In this structure, employees themselves design and do work accordingly with accountability for their results and performance. Companies like Google, Amazon, and HP extensively use this structure to increase productivity.

1.7.3   SELF-MANAGED WORK TEAMS

Self-managed teams or self-directed, self-regulating, or high-performance work teams, these work designs consist of members performing interrelated tasks. Self-managed teams typically are responsible for a complete product or service or a major part of a larger production process. They control member's task behavior and make decisions about task assignments and work methods. Self-managed work teams are being implemented at a rapid rate across a range of industries and organizations such as Intel, General Electric, and Motorola. A 2001 survey of Fortune 1000 companies found that 70% of these firms were using self-managed work teams, a nearly 50% increase from 1987.

1.7.4   LEARNING ORGANIZATIONS

Garvin (1993) defines learning organization as an organization skilled at creating, acquiring, and transferring knowledge and at modifying its behavior to reflect new knowledge and insights.

1.7.5   VIRTUAL ORGANIZATIONS

A variation in the network structure has been referred to as virtual organization. A virtual organization tends to coordinate and link people from many different locations to communicate and take decisions on real time issues (Markus et al., 2000). Generally these organizations exist temporarily and have the latest type of structure.

1.8   CHALLENGES FOR TODAY'S ORGANIZATIONAL DESIGN

- Knowledge management
- Building learning organization
- Keeping employees connected
- Managing global structural issues

1.9   ORGANIZATIONAL LIFE CYCLE

How some organizations survive and prosper for long time and mostly fail and die? Researchers have suggested that organizations experience a predictable sequence of stages of growth and change: the organizational life cycle. The four main stages of the organizational life cycle are birth, growth, decline, and death; organizations pass through these stages as they grow and evolve. Some of the companies go directly from birth to death. Population ecology theory states that organizational birthrate in a new environment is very high at first but tapers off as the number of successful organizations in a population increases. According to Greiner's five stage model of organizational growth, organizations experience growth through five stages:

Stage 1: Growth through creativity
Stage 2: Growth through direction

Stage 3: Growth through Delegation
Stage 4: Growth through Coordination
Stage 5: Growth through Collaboration

The liability of newness, threatens young organizations, and the failure to develop a stable structure can cause early decline in an organization's ability to obtain resources from its stakeholders. Decline sometimes occurs as organizations grow too fast or too much. Organizational Decline is the stage that an organization enters when it fails to anticipate, recognize, avoid, neutralize, or adapt to external or internal pressures that threaten its long term survival. Factors that precipitate decline of organization include organizational inertia and changes in the environment. Weitzel and Jonsson have identified five stages of decline: (a) blinded, (b) inaction, (c) faulty action, (d) crisis, and (e) dissolution. Managers can turn the organization around at every stage except the dissolution stage.

A new organization can enhance its legitimacy by choosing the goals, structure, and culture that are used by other successful organizations in theirpopulations.

1.10 ORGANIZATIONAL CULTURE

Individuals get affected by the culture, in which they live. In families members are taught certain norms, behavior by the process of socialization. The same process is true for the organizations. Every organization has its own vision and mission which can be seen or communicated through policies and procedures. Hagbergand Heifetz (2000) concede culture drives the organization and its actions. It is somewhat like the operating system of the organization. It guides how employees think, act, and feel. According to Chatman and Jehn (1994), elements of culture includes innovation, stability, orientation toward people, result orientation, easygoingness, and attention to detail and collaborative orientation. Functions of organizational culture include:

- Culture gives a sense of identity to its members
- Culture helps to generate commitment among employees
- Culture serves to clarify and reinforce standards of behavior

Every organization has its own culture; Sonnenfeld (1988) notes that these commonly fall into one of the four different categories–the academy, the club, the baseball team, and the fortress.

1.10.1 TRANSMITTING THE ORGANIZATIONAL CULTURE

1.10.1.1 ROLE OF LEADERS

Leaders play a crucial role in shaping, reinforcing, and transmitting the culture of an organization (Schein, 1983). The leader communicates the culture to employees by being consistent in what they do, and therefore sends signals about their expectations of the employees.

1.10.1.2 SOCIALIZATION

Employees are socialized into organizations, just as people are socialized into families and societies. Socialization is a process through which children learn to be adults in a society–how they learn, what is acceptable and polite behavior and what is not, interaction, communication, and so on. Research into the process of socialization indicates that for many employees, socialization programs do not necessarily change their values but make them more aware of the differences between personal and organizational values and help them develop ways to cope with the differences(Hebden, 1986).

1.10.1.3 PROCESS OF SELECTION

At the time of selection, certain criteria are chosen to select the most appropriate person for the job. This is the first step toward choosing the right organization by the applicant, thereby initiating the process of learning the culture of the organization. In the next stage, when the applicant is face to face in the form of interview, the organization decides to either select or reject the person. Consciously or subconsciously, the decision taken to select the person is based on the matching of the value system of the person with the value system (culture) of the organization.

1.11   IS IT POSSIBLE TO CHANGE CORPORATE CULTURE?

Culture is as deep seated in an organization as personality is in an individual, so it is quite difficult to change the culture of an organization. While changing the culture, the depth of cultural change has to be identified. Cultural change may involve certain issues of ethical and legal sense. Change is essential when business environment like external forces of technology, economic and political changes is going on, or in case of mergers, acquisitions, and an organization planning for expansion.

Managers in today's organizations have to organize processes, and think internationally considering people of all nationalities. The culture of countries is studied on five dimensions: power distance, uncertainty avoidance, individualism-collectivism, masculinity-feminity, and long term. For managers to be efficient internationally requires intuition, to be socially perceptive and adaptive.

KEYWORDS

- **organizational goals**
- **continuous improvement**
- **decision-making process**
- **organizational structure**
- **self-management**

CHAPTER 2

LEADERSHIP

CONTENTS

2.1   HOW DO WE DEFINE LEADERSHIP?

Leadership is a process by which a person influences others to accomplish an objective and directs the organization in a way that makes it more cohesive and coherent. Leaders carry out this process by applying their leadership attributes, such as beliefs, values, ethics, character, knowledge, and skills.

Are you aware that up to half of all corporate decisions may be wrong? An organizational psychologist examined 356 decisions made by managers in companies of medium to large size in the United States and Canada and showed that these actions were directly related to the manager's personal style of leadership (Nutt, 1999).

Zipkin (2000) found in an opinion poll of several hundred workers that only 11% of those who rated their supervisor's performance as excellent said they would be seeking to change jobs in the coming year. In contrast, 40% of those who rated their supervisor's performance as poor said they would be looking for another job.

The quality of leadership is a critical factor in the workplace. Organizations are greatly concerned with selecting, developing, and supporting their managers and executives, and in making the best use of their leadership abilities on the job.

2.1.1   LEADERSHIP DEFINED

Leadership has been defined in many different ways, Stogdill (1974, p.259) concluded that "there are almost as many definitions of leadership as there are persons who have attempted to define the concept." We see leadership has been defined in terms of traits, behaviors, influence, interaction patterns, role relationships, and occupation of an administrative position. The following definitions are presented over the past 50 years:

"Behavior of an individual…directing the activities of a group toward a shared goal." (Hemphill & Coons, 1957, p.7)

"It is the influential increment over and above mechanical compliance with the routine directives of the organization." (Katz & Kahn, 1978, p.528)

"Leadership is about articulating visions, embodying values, and creating the environment within which things can be accomplished." (Richards & Engle, 1986, p.206)

"Leadership is the process of making sense of what people are doing together so that the people will understand and be committed." (Drath & Palus, 1994, p.4)

"The ability of an individual to influence, motivate and enable others to contribute toward the effectiveness and success of the organization...." (House et al., 1999, p.184)

Definitions reflect the assumption that it involves a process whereby intentional influence is exerted over other people to guide, structure, and facilitate activities and relationships in a group or organization.

Guru of leadership Warren Bennis gave the title "The End of Leadership" to make the point that effective leadership cannot exist without the full inclusion, initiatives, and the cooperation of employees. One cannot become a great leader without great followers. Barry Posner, another leadership guru, has following observations about leadership:

"In the past, business believed that a leader need to be human. They need to be in touch, they need to be empathetic, and they need to be with people. Leader need to be a part of what's going on, not apart from what's going on."

Technological advancement brought globalization that has changed the traditional view of leaders as the heroic individual, whose positional power, intellectual strength, and persuasive gifts motivate followers. But this is not necessarily the ideal in Asia, nor does it match the requirements in large global corporations, where forms of distributed and shared leadership are needed to address complex interlocking problems.

A Gallup survey suggests that most of the employees think that it is the leader, not the company that guides the culture and creates situations where workers can be happy and successful. Although many specific definitions could be cited, but major would be interpreted on the theoretical orientation taken. Besides influence, leadership has been defined in terms of group processes, personality, compliance, particular behaviors, persuasion, power, goal achievement, interaction, role differentiation, initiation of structure, and combinations of two or more of these. Bennis and Thomas concluded:

One of the most reliable indicators and predictors of true leadership is an individual's ability to find meaning in negative events and to learn from even the most trying circumstances. Put another way, the skills required to conquer adversity and emerge stronger and more committed than ever are the same ones that make for extraordinary leaders.

Avolio, Luthans, and colleagues at the Leadership Institute at the University of Nebraska concentrate on authentic leaders, who say,

To know oneself, to be consistent with oneself, and to have a positive and strength-based orientation toward one's development and the development of others. Such leaders are transparent with their values and beliefs. They are honest with themselves and others. They exhibit a higher level of moral reasoning capacity, allowing them to judge between gray and shades of gray.

A simple definition in Fortune article states: "When you boil it all down, contemporary leadership seems to be a matter of aligning people toward common goals and empowering them to take the actions needed to reach them."

2.1.2 STUDIES OF LEADERSHIP

There are numerous studies on leadership unlike other concepts of organizational behavior.

2.1.2.1 THE IOWA LEADERSHIP STUDIES

In the late 1930s, Ronald Lippitt and Ralph K. White conducted studies on leadership with the guidance of Kurt Lewin at the University of Iowa. Lewin is the father of group dynamics. He explored three leadership styles to find which was the most effective:the autocratic style, the democratic style, and finally the laissez-faire style described the leader who let the group members make decisions on their own and complete the work in whatever way it saw fit. Unfortunately, the question of the most effective leadership was not found. Later studies of the autocratic and democratic styles showed mixed results. In some instances it was found that democratic styles produced higher performance levels than the autocratic style and vice versa. Group members were more satisfied under a democratic leader than under an autocratic one.

2.1.2.2   THE OHIO STATE LEADERSHIP STUDIES

At Ohio State University the Bureau of Business Research initiated a series of studies on leadership. Studies were done on Air Force commanders, members of bomber crews; officers, noncommissioned personnel, civil administrators in the Navy Department; manufacturing supervisors; executives of regional cooperatives; college administrators; teachers, principals, and leaders of various student and civilian groups. It was found that no satisfactory definition of leadership existed.

The Ohio State studies were the first to point out and emphasize the importance of both task and human dimensions for assessing leadership. This two dimensional approach made less gap in task orientation of the scientific management movement and the human relations emphasis. Today leadership is recognized as both multidimensional, as first pointed out by the Ohio State studies, and multilevel (person, dyad, group, and collective/community).

2.1.2.3   THE EARLY MICHIGAN LEADERSHIP STUDIES

Almost about the same time period when Ohio State studies were being conducted, a researcher group from the Survey Research Centre at the University of Michigan began their studies of leadership. At the Prudential Insurance Company, 12 high–low productivity pairs were selected for study. Each pair comprised of a high producing section and a low producing section, with variables, such as type of work, conditions, and methods being the same in each pair. Results revealed that supervisors of high producing sections were significantly more likely to be general rather than close in their supervisory styles and be employee centered. The low producing section supervisors had opposite characteristics and techniques. In this study, it was found that employee satisfaction was not directly related to productivity, the type of supervision was the key to their performance.

2.2   ARE LEADERS BORN OR MADE?

Effective leaders are not simply born or made, they are born with some leadership ability and develop it. So, natural ability may offer advantages or disadvantages to a leader.

2.3   LEADER VERSUS MANAGERS

The terms management and leadership are often used almost interchange-ably, but not every manager is a leader and not every leader is a manager. The authority that a manager has to set goals, develop plans, and assign work makes that person a boss, not a leader. Leaders have personal qualities that motivates others to follow their direction and work toward a common goal. Warren Bennis, business professor at the University of South Cali-fornia, says the primary difference is that leaders are concerned with doing the right thing, while managers are concerned with doing things right. Another view is that while managers are concerned with control, leaders are more concerned with expanding people's choices.

Leaders	Managers
Acts as a transactor	Act as a transformer
The leaders are responsible for the identification of goals, initiation of the development of a vision for the unleash-ing of the requisite energy	Manager's task is to procure, allocate, and coordinate the requisite material and man-power according to rules
	Their vision is the shared vision
Ability to create a vision	
Maintains	Develops
Accepts the status quo	Challenges the status quo
Administers	Innovates

"Managers solve problems so that others can do their work, while leaders inspire and motivate others to find their own solutions"–CHUCK WILLIAMS

2.4   QUALITIES OF LEADERS

What is that separates a leader from a follower? Some of the attributes commonly accepted as essential for leaders are listed in Table 1.

TABLE 1:   Essential Leadership Qualities

Ambition	Persistence
Creativity	High energy
Assertiveness	Decisiveness
Self-confidence	Persuasiveness
Emotional intelligence	Integrity

2.5 THE PERSONALITY PROFILE OF EFFECTIVE LEADERS

Effective leaders have specific personality traits. Motivation is another aspect of personality related to leadership effectiveness. The classic research of McClelland and his colleagues (e.g., McClelland & Boyatzis, 1982) identified three leader motives: need for power, need for achievement, and need for affiliation.

2.5.1 ACHIEVEMENT MOTIVATION THEORY

It explains and predicts behavior and performance based on a person's need for achievement, power, and affiliation. David McClelland originally developed Achievement Motivation Theory in the 1940s. According to this theory, we have needs and needs motivates us to satisfy them. The human behavior is motivated by their needs. This is an unconscious process. All humans possess the need for achievement, power, and affiliation, but to varying degree.

2.5.2 THE NEED FOR ACHIEVEMENT (N-ACH)

It is the unconscious concern for excellence in accomplishments through individual efforts. Individuals having strong N-Ach is usually having internal locus of control, high energy, and self-confidence. They want challenge, excellence, take calculated risks, desire concrete feedback on their performance, and work hard. McClelland's research showed that only about 10% of the US population has a "strong" dominant need for achievement. There is evidence of a correlation between high achievement need and high performance in the general population. People with high N-Ach tend to enjoy entrepreneurial type positions.

2.5.3 THE NEED FOR POWER (NPOW)

It is the unconscious concern for influencing others and seeking positions of authority. These people are more concerned about getting their own way than about what others think of them. High nPow is categorized as the

Big Five dimension of surgency. They are attuned to power and politics as essential for successful leadership.

2.5.4   THE NEED FOR AFFILIATION (N-AFFIL)

The need for affiliation is the unconcious concern for developing, maintaining and restoring close personal relationships. People with strong N-Affil have the trait of sensitivity to others. High N-Affil is categorized as the Big Five dimension of agreeableness. People with high N-Affil tend to seek close relationships with others, wanting to be liked by others, enjoying social activities, and seeking to belong. They lay importance to friends and relationships.

2.6   LEADERSHIP ATTITUDES

Attitudes are positive or negative feelings about people, things, and issues. There has been considerable interest in how attitudes affect performance and companies are recruiting workers with positive attitudes. Optimism is a good predictor of job performance. W. Marriot, Jr., President of Marriot Corporation, stated that the company's success depends more upon employee attitudes than any other single factor. People with positive, optimistic attitudes generally have a well-adjusted personality profile, and successful leaders have positive, optimistic attitudes. So follow Jack Welch's advice and have a positive attitude to advance. Like personality traits, attitudes have an important influence on behavior and performance.

2.6.1   THEORY X AND THEORY Y

Theory X and Theory Y attempt to explain and predict leadership behavior and performance based on leader's attitude about followers. Douglas McGregor classified attitudes or belief systems, which he called assumptions, as Theory X and Theory Y. People with theory X attitudes hold that employees dislike work and must be closely supervised in order to do their work. Theory Y attitudes hold that employees like to work and do not need to be closely supervised in order to do their work. Managers with Theory X attitudes tend to have a negative, pessimistic view of employees and display more coercive, autocratic leadership styles using external means of controls such as threats and punishment. Managers with Theory Y attitudes tend to have a positive, optimistic view of employees and display more participative leadership styles using internal motivation and rewards. Managers should acknowledge the influence of attitudes on behavior and performance. A study of over 12,000 managers explored the relationship between managerial achievement and attitudes toward subordinates.

2.6.2   THE PYGMALION EFFECT

Research by J. Sterling Livingston popularized this theory, which proposes that leaders' attitudes toward and expectations of followers and their treatment of them explain and predict followers' behavior and performance.

In a study of welding students, the foreman who was training the group was given the names of students who were quite intelligent and would do well. Actually, the students were selected at random. The only difference was the foreman's expectations. The so called intelligent students did significantly outperform the other group members. Why this happened is what this theory is all about: The foreman's expectations influenced the behavior and performance of the followers.

2.6.3   SELF-CONCEPT

Self-concept refers to the individual's positive or negative attitudes about himself or herself. If you have a positive view of yourself as being a capable person, you will tend to have the positive self-confidence trait. A related concept, self-efficacy is the belief in your own capability to perform in a specific situation. It is discussed in detail in Chapter 6. Successful leaders have positive attitudes with strong self-concepts, are optimistic and believe they can make a positive difference.

2.6.4   HOW TO BUILD A POSITIVE ATTITUDE AND SELF-CONCEPT

1. Accept compliments
2. Focus on strengths

3. Avoid negative people
4. Set and achieve goals
5. Exercise/yoga/or some spiritual discipline for managing stress

When things go wrong and you are feeling low, do something to help those who are worse off than you

2.6.5 HOW ATTITUDES DEVELOP LEADERSHIP STYLES CAN BE SEEN FROM THIS EXHIBIT 2.2

	Theory Y attitudes	Theory X attitudes
Positive self-concept	Leader gives and accepts positive feedback, expects others to succeed	The leader is bossy, impatient, pushy, praising less, and more criticizing, is autocratic
Negative self-concept	Leader is afraid to make decisions, is unassertive, and self-blaming when things go wrong	The leader typically blames others when things go wrong, is pessimistic, and promotes a feeling of hopelessness among followers

2.7 ETHICS AND LEADERSHIP

Ethics is an especially hot topic, because it is a major concern to both managers and employees. Ethics are the standards of right and wrong that influence behavior.

Attitudes precede action. Leadership attitudes are ethical values. How-to guides without right attitudes are empty gestures. Doing right and doing well are not mutually exclusive; in fact, doing right is an integral pan of doing well. But it is also possible to place too much emphasis on ethics to the detriment of other important facet of leadership. Balance is the key. Because the word "ethics" means different things to different people, its definition should be spelled out and not left open for interpretation. The drafters must remember that these policies are not carved in stone, time and experience may reveal detrimental unintended consequence. Amending the ethics statement is acceptable as long as it remains true to the original spirit and intent.

2.7.1 THE ROLE OF LEADERSHIP

The catalyst for the implementation of the solution lies with the leadership of the organization supporting those that behave ethically, both financially and in other ways. Every conscientious businessperson should make it a priority to explore ethical behavior, and learn how to make ethical decisions.

Warren Bennie, business professional and founding chairman of the Leadership Institute at the University of Southern California Marshall School of Business, said that, "Exemplary leaders create a climate of candor throughout their organizations. They remove the organizational barriers and the fear–that cause people to keep bad news from the boss. They understand that this closest to customers usually have the solutions but can do little unless a climate of candy allows problems to be discussed. Exemplary leaders share information about what is going on in the organization, the industry, and the world; and they treat candor as one measure of personal and organizational performance.

Character involves not just doing the right thing but doing the right thing for the right reasons. When the leaders walk the talk and fight the good fight, then the organization observes and begins to follow. The first step, then, to becoming whole lies in courageous decisions to be open-minded, to make self-transcending commitments, and to help create a common culture.

2.7.2 FACTORS AFFECTING ETHICAL BEHAVIOR

2.7.2.1 PERSONALITY TRAITS AND ATTITUDES

Ethical behavior is linked to individual's needs and personality. Leaders with surgency (dominance) personality traits have two choices: to use power for personal benefit or to use socialized power. To gain power and to be conscientious with high achievement, some people will use ethical behavior. People who are open to new experiences are often ethical. In a recent survey, it was found that over two-thirds (71%) of Americans rated corporations low for operating in a fair and honest manner. It has been said that a culture of lying is infecting American business.

Some of the reasons for lack of ethics are that many people have the idea that if you are making money, any behavior is acceptable. This notion is supported by reality TV shows set in cutthroat world where business competition is likened to a struggle for survival on an island. The message people are absorbing from these shows is that success means clawing your way to the top of the heap at the expense of your competition, as long as you donot get caught. Being unethical is a more accepted part of doing business today, unfortunately, being ethical is not often rewarded, and being unethical is even rewarded. A survey found that 19% of employees have seen coworkers steal from customers, vendors, or the public and that 12% have seen coworkers steal from customers or the company. And 35% admit keeping quiet when they see coworkers' misconduct. People tend to make rapid judgments about ethical dilemmas. So slow down your decisions that affect various stakeholders. Seek out mentors who can advise you on ethical dilemmas.

A few ways by which we can find courage to do the right thing are following:

- Draw strength from others.
- Take risk without fear of failure.
- Channelize you frustration and anger for good.
- Look at a larger picture/higher purpose.

One of the examples of ethical leadership includes the event when Warren Buffet's taking over Salomon Brothers it was full of scandals for unethical behavior. Buffet called a meeting with employees saying the unethical behavior had to stop. He was the compliance officer; he gave his home phone number and told employees to call him if anyone observed any unethical behavior. What makes a person want to follow a leader? People want to be guided by those they respect and who have a clear sense of direction. To gain respect, they must be ethical. A sense of direction is achieved by conveying a strong vision of the future.

2.8　THE ROLE OF LEADERSHIP IN ORGANIZATION

In an organization where there is faith in the abilities of formal leaders, employees will look toward the leaders for a number of things. During drastic change times, employees will expect effective and sensible planning, confident and effective decision-making, and regular, complete communication that is timely. Also during these times of change, employees will perceive leadership as supportive, concerned, and committed to their welfare, while at the same time recognizing that tough decisions need to be made. The best way to summarize is that there is a climate of trust between leader and the rest of the team. The existence of this trust brings hope for better times in the future, and that makes coping with drastic change much easier.

In organizations characterized by poor leadership, employees expect nothing positive. In a climate of distrust, employees learn that leaders will act in indecipherable ways and in ways that do not seem to be in anyone's best interests. Poor leadership means an absence of hope, which, if allowed to go on for too long, results in an organization becoming completely nonfunctioning. The organization must deal with the practical impact of unpleasant change, but more importantly, must labor under the weight of employees who have given up, have no faith in the system, or in the ability of leaders to turn organization around.

Leadership before, during, and after change implementation is the key to getting through the swamp. Unfortunately, you have not established a track record of effective leadership, by the time you have to deal with difficult changes, it may be too late.

Leaders play a critical role during change implementation, the period from the announcement of change through the installation of the change. During this middle period the organization is the most unstable, characterized by confusion, fear, loss of direction, reduced productivity, and lack of clarity about direction and mandate. It can be a period of emotionalism, with employees grieving for what is lost, and initially unable to look to the future.

During this period, effective leaders need to focus on two things. First, the feelings and confusion of employees must be acknowledged and validated. Second, the leader must work with employees to begin creating a new vision of the altered workplace, and helping employees to understand the direction of the future. Focusing only on feelings, may result in wallowing. That is why it is necessary to begin the movement into the new ways or situations. Focusing only on the new vision may result in the perception that the leader is out of touch, cold, and uncaring. A key part of leadership in this phase is knowing when to focus on the pain, and when to focus on building and moving into the future.

The two most important keys to effective leadership:

In Hay's study examined over 75 key components of employee satisfaction. They found:

- Trust and confidence in top leadership was the single most reliable predictor of employee satisfaction in an organization.
- Effective communication by leadership in three critical areas was the key to winning organizational trust and confidence:

    1. Helping employees understand the company's overall business strategy.
    2. Helping employees understand how they contribute to achieving key business objectives.
    3. Sharing information with employees on both how the company is doing and how an employee's own division is doing –relative to strategic business objectives.

So in a nutshell, a leader must be trustworthy and able to communicate a vision of where the organization needs to go.

The road to great leadership (Kouzes & Posner, 1987) that is common to successful leaders:

- Challenge the process–First, find a process that you believe needs to be improved the most
- Inspire a shared vision–Next, share your vision in words that can be understood by your followers
- Enable others to act –Give them the tools and methods to solve the problem
- Model the way–When the process gets tough, get your hands dirty. A boss tells others what to do….a leader shows that it can be done
- Encourage the heart–Share the glory with your followers' heart, while keeping the pains within your own

2.9 LEADERSHIP THEORIES

2.9.1 THEORETICAL APPROACHES TO LEADERSHIP

Various theoretical approaches have developed from time to time to explain leadership. There are several distinct bases of theory for leadership.

2.9.2 THE TRAIT APPROACH

The trait approach emphasizes the personal attributes of leaders. Advances in trait research led to a change of focus from abstract personality traits to specific attributes that can be related directly to behaviors required for effective leadership in a particular situation. This more directed approach revealed that some traits increase the likelihood of success as a leader, even though none of the traits guarantees success (Kirkpatrick & Locke, 1991). Some of the individual traits related to leadership success are high energy level, tolerance for stress, emotional maturity, integrity, and self-confidence.

The six traits which identify ways in which leaders differ from nonleaders are–drive the desire to lead (wanting to influence and will-ingness to take responsibility), honesty and integrity, self-confidence, intelligence, and job-relevant knowledge. Yet possession of these traits is no guarantee of leadership because situational factors cannot be ignored (Robbins & Judge, 2009).

It is a topic of constant study and discussion where everyone seems to have a view and where definitions of leadership are as varied as the expla-nations. More recent day definitions and study have focused on leadership and change, vision building, and empowering others.

Hawthorne studies and Kurt Lewin and Likert participative styles of leadership lead to increased job satisfaction and higher performance.

Participative leadership can be regarded as a distinct type of behavior, although it may be used in conjunction with specific task and relationbe-haviors (Likert, 1967; Yukl, 1971).

Instrumental theories stress task- and person-oriented behavior (e.g., participation and delegation) by the leader to gain effective performance from others.

Inspirational theories include charismatic leaders, transformational leadership. The leader appeals to values and vision, and enthuses others raising confidence in others and motivating them for change.

Informal leadership looks at behaviors associated with those who are not appointed to authority but assume leadership in other ways.

Path–goal theory looks at what leaders must do to motivate people to perform well and to get satisfaction from work. It draws on the expectancy theory of motivation–four leadership styles: supportive, directive, partici-pative, and achievement oriented.

The choice of style depends on the task and the individual, for example, routine tasks = supportive style, complex = directive leadership.

Briefly the summarized styles are:

Directive leadership–There is no participation from subordinates and associates who know exactly what is expected of them and leader gives directions which are specific.

Supportive leadership–The leader is approachable and shows genuine concern for associates.

Participative leadership–The leader asks for and takes suggestions from associates but still takes decision.

Achievement-oriented leadership–The leader sets challenging goals for subordinates and shows confidence that they will attain these goals.

The path–goal theory and how it differs in one respect from Fiedler's contingency modelsuggests that these various styles can be, and actually are used by the same leader in different situations.

In relation to the first situational factor, the theory asserts: Leader's behavior will be acceptable to subordinates to the extent that the subordinates see such behavior as either an immediate source of satisfaction or as instrumental to future satisfaction. With respect to second situational factor, the theory states:

Leader behavior will be motivational to the extent that (1) it makes satisfaction of subordinate needs contingent on effective performance and (2) it complements the environment of subordinates by providing the coaching, guidance, support, and rewards which are necessary for effective performance and which may otherwise be lacking in subordinates or in their environment.

We should take consideration of these theories origin which is North American and do not necessarily take account of cross-cultural differences. They are also almost all drawn from observation.

Fielder, one of the leaders of the contingency school, offered a continuum ranging from task-focused to people-focused leadership. He argued that the most effective style depended on the quality of relationships, relative power position between the leader and the led, and the nature of the task. He also argued that the style adopted is relatively stable and a feature of a leader's personality and could therefore be predicted. He distinguishes between task-oriented and relation-oriented leaders. Situational favorableness was explained by Fiedler in terms of three empirically derived dimensions:

1. The leader–member relationship which is critical in determining the situation's favorableness
2. The degree of task structure which is the second most important input into favorableness of the situation
3. The leader's position power obtained by formal authority

Situations are favorable for the leader if all three of these dimensions are high.

Hersey and Blanchard–situational leadership where dimensions are linked to task and relational behavior. Task behavior focuses on defining roles and responsibilities, whereas relational behavior is more about providing support to teams. The extent to which either is used depends on the person's job maturity and psychological security. Their test looks at elements around delegate, participate, sell, or tell.

2.9.3 THEORIES OF LEADERSHIP STYLE

Fundamental to the management of people is an understanding of the importance of leadership. Managers must lead, and as such must accept responsibility for the activities and successes of their departments. All leaders must exercise authority, but leadership style will vary. It is generally accepted that a leader's style will affect the motivation, efficiency, and effectiveness of their employees.

The main leadership theories present two basic approaches–task-centered and employee centered. Tannenbaum and Schmidt suggest that leadership style is a continuum, and that the appropriate style depends on the characteristics of the leader, the subordinates, and of the situation.

In a more contemporary approach, known as "action-centered leadership," John Adair suggests that there are three basic needs that result in differing leadership styles: the needs of the task, the needs of the group, and the needs of the individual. Feidler, on the other hand, takes a more psychological approach to defining leadership.

As these approaches to leadership vary, it is interesting to explore the differences. Tannenbaum and Schmidts continuum-based theory suggests a range of styles ranging from autocratic to democratic, although not suggesting that any one style within the continuum is right or wrong.

At one end of the continuum is the dictatorial style–the manager makes decisions and enforces them (the so-called tells approach) or, in a slightly gentler way, "sells" their decision (the tells and sells approach). Further along the continuum, is the autocratic style, where the manager suggests ideas and asks for comments (the tells and talks approach), or the manager presents outline ideas, seeks comments, and amends the ideas accordingly (the consults approach).

The next step in the continuum is the democratic approach. Here the manager presents a problem, again seeks ideas and makes a decision (the involves approach), or allows employees to discuss the issue and make a decision (the delegates approach). Finally, the continuum ends with the laissez-faire approach. Here the manager allows employees to act in whichever way they wish, within specified limits (the abdicates approach).

However, Tannenbaum and Schmidts continuum is not a static model. It recognizes that appropriate style depends on both, the leader's personality, values, and natural style, and the employees' knowledge, experience, and attitude. Furthermore, the range of situations which present themselves to a leader depend on factors, such as the culture of the organization, time pressure, the amount of authority and the amount of responsibility the leader has. This last factor is dependent–as is so often the ease–upon the organization's general environment.

A more contemporary approach is to regard leadership as comprising a number of different skills (action-centered leadership), a theory associated with the writer John Adair. This idea recognizes that leadership style is determined by three interrelated variables: the needs of the task, the needs of the group, and the needs of the individual. The leader needs to balance the relative importance of all three, with emphasis given to identifying and acting upon the immediate priority.

In contrast, a different, more psychological approach to leadership, described by the writer Feidler, suggests a relationship between leadership styles and departmental effectiveness and success. He distinguishes between two types of leaders– those who are psychologically close and those who are psychologically distant. Psychologically close managers prefer informal relationships, are sometimes over concerned with human relations, and favor informal rather than formal contacts. This is sometimes called "relationship oriented." Psychologically distant managers prefer formal relationships. They tend to be reserved in their personal relationships even though they often have good interpersonal skills. This approach is sometimes called "task oriented."It is, of course, vital to recognize that no leadership style is correct, and that style is always dependent upon the particular situation, and the nature and culture of the organization.

"The task" refers to the setting of objectives for the department, planning and initiating the task, allocating responsibilities, setting and verifying performance standards, and establishing a control system.

"Group needs" require team building so that mutual support and understanding is achieved, standards established, training provided and most importantly, communication and most importantly, communication and information channels opened.

"Individual needs" recognize the development and nurturing of individual achievement, of motivation, the encouragement of creativity, delegation of authority to encourage group support, and attention to any problems or issues.

2.9.3.1 SITUATIONAL LEADERSHIP

ASSUMPTIONS

The best action of the leader depends on a range of situational factors.

STYLE

When a decision is needed, an effective leader does not just fall into a single preferred style, such as using transactional or transformational methods. In practice, as they say, things are not that simple.

Factors that affect situational decisions include motivation and capability of followers. This, in turn, is affected by factors within the particular situation. The relationship between followers and the leader may be another factor that affects a leader's behavior as much as it does the follower's behavior.

The leaders' perception of the follower and the situation will affect what they do rather than the truth of the situation. The leader's perception of themselves and other factors such as stress and mood will also modify the leaders' behavior.

Yukl (1989) seeks to combine other approaches and identifies six variables:

- Subordinate effort: the motivation and actual effort expended
- Subordinate ability and role clarity: followers knowing what to do and how to do it
- Organization of the work: the structure of the work and utilization of resources
- Cooperation and cohesiveness: of the group in working together
- Resources and support: the availability of tools, materials, people, etc.
- External coordination: the need to collaborate with other groups

Leaders here work on such factors as external relationships, acquisition of resources, managing demands on the group, and managing the structures and culture of the group.

DISCUSSION

Tannenbaum and Schmidt (1958) identified three forces that led to the leader's action: the forces in the situation, the forces in then follower, and also forces in the leader. This recognizes that the leader's style is highly variable, and even such distant events as a family argument can lead to the displacement activity of a more aggressive stance in an argument than usual.

Maier (1963) noted that leaders not only consider the likelihood of a follower accepting a suggestion, but also the overall importance of getting things done. Thus in critical situations, a leader is more likely to be directive in style simply because of the implications of failure.

2.9.3.2 CONTINGENCY THEORY

Contingency means "it depends." One thing depends on other things, and for a leader to be effective there must be an appropriate fit between the leader's behavior and style, and the followers, and the situation. Different groups also prefer different leadership styles. Leaders display a range of behavior in different situations, because leadership is largely shaped by contextual factors that not only set the boundaries within which leaders and followers interact but also determine the demands and constraints confronting the leader. Contingency theories basically all argue that the "right" or an effective leadership style varies according to the context. For example, Blake and Mouton's managerial grid which has been very influential in organization development practice.

ASSUMPTIONS

The leader's ability to lead is contingent upon various situational factors, including the leader's preferred style, the capabilities and behaviors of followers, and also various other situational factors.

DESCRIPTION

Contingency theories are a class of behavioral theory, which contends that there is no one best way of leading and that a leadership style that is effective in some situations may not be successful in others.

An effect of this is that leaders who are very effective at one place and time may become unsuccessful either when transplanted to another situation or when the factors around them change.

This helps to explain how some leaders who seem for a while to have the "Midas touch" suddenly appear to go off the boil and make very unsuccessful decisions.

DISCUSSION

Contingency theory is similar to situational theory in that there is an assumption of no simple one right way. The main difference is that situational theory tends to focus more on the behaviors that the leader should adopt, given situational factors (often about follower behavior), whereas contingency theory takes a broader view that includes contingent factors about leader capability and other variables within the situation. According to Fiedler, relationships, task structure, and power are the three key factors that drive effective leadership styles.

2.9.3.3   TRANSACTIONAL LEADERSHIP

ASSUMPTIONS

- People are motivated by reward and punishment.
- Social systems work best with a clear chain of command.
- When people have agreed to do a job, a pan of the deal is that they cede all authority to their manager.
- The prime purpose of a subordinate is to do what their manager tells them to do.

STYLE

The transactional leader works through creating clear structures whereby it is clear what is required of their subordinates, and the rewards that they get for following orders. Punishments are not always mentioned, but they are also well-understood and formal systems of discipline are usually in place.

The early stage of transactional leadership is in negotiating the contract whereby the subordinate is given a salary and other benefits, and the company (and by implication the subordinate's manager) gets authority over the subordinate.

When the transactional leader allocates work to a subordinate, they are considered to be fully responsible for it, whether or not they have the resources or capability to carry it out. When things go wrong, then the subordinate is considered to be personally at fault, and is punished for their failure (just as they are rewarded for succeeding). Whereas, transformational leadership has more of a 'selling' style, transactional leadership, once the contract is in place, takes a "telling" style.

DISCUSSION

Despite much research that highlights its limitations, transactional leadership is still a popular approach with many managers. Indeed, in the leadership versusmanagement spectrum, it is very much toward the management end of the scale.

The main limitation is the assumption of "rational man," a person who is largely motivated by money and simple reward, and hence whose behavior is predictable. The underlying psychology is behaviorism, including the Classical Conditioning of Pavlov and Skinner's Operant Conditioning. These theories are largely based on controlled laboratory experiments (often with animals) and ignore complex emotional factors and social values.

In practice, there is sufficient truth in behaviorism to sustain transactional approaches. This is reinforced by the supply-and-demand situation of employment, coupled with the effects of deeper needs, as in Maslow's Hierarchy. When the demand for a skill outstrips the supply, then transactional leadership often is insufficient, and other approaches are more effective.

2.9.3.4   PARTICIPATIVE LEADERSHIP

ASSUMPTIONS

- Involvement in decision-making improves the understanding of the issues involved by those who must carry out the decisions.
- People are more committed to actions where they have been involved in the relevant decision-making.
- People are less competitive and more collaborative when they are working on joint goals.
- When people make decisions together, the social commitment to one another is greater, and thus, increases their commitment to the decision.
- Several people deciding together make better decisions than one person alone.

STYLE

A participative leader, rather than taking autocratic decisions, seeks to involve other people in the process, possibly including subordinates, peers, superiors, and other stakeholders. Often, however, as it is within the managers' whom to give or deny control to his or her subordinates, most participative activity is within the immediate team. The question of how much influence others are given, thus may vary on the managers preferences and beliefs, and a whole spectrum of participation is possible, as in the table below.

< Not participative			Highly participative >	
Autocratic decision by leader	Leader proposes decision, listens to feedback, then decides	Team proposes decision, leader has final decision	Joint decision with team as equals	Full delega-tion of deci-sion to team

There are many varieties on this spectrum, including stages where the leader sells the idea to the team. Another variant is for the leader to describe the "what" of objectives or goals and let the team or individuals decide the "how" of the process by which the "how" will be achieved (this is often called "Management by Objectives").

The level of participation may also depend on the type of decision being made. Decisions on how to implement goals may be highly partici-pative, while decisions during subordinate performance evaluations are more likely to be taken by the manager.

This approach is also known as consultation, empowerment, joint deci-sion-making, democratic leadership, management by objectives (MBO) and power sharing.

Participative leadership can be a sham when managers ask for opinions and then ignore them. This is likely to lead to cynicism and feelings of betrayal.

2.9.3.5   TRANSFORMATIONAL LEADERSHIP

ASSUMPTIONS

- People will follow a person who inspires them.
- A person with vision and passion can achieve great things.
- The way to get things done is by injecting enthusiasm and energy.

STYLE

Working for a transformational leader can be a wonderful and uplifting experience. They put passion and energy into everything. They care about you and want you to succeed.

Developing the vision

Transformational leadership starts with the development of a vision, a view of the future that will excite and convert potential followers. This vision may be developed by the leader, by the senior team or may emerge from a broad series of discussions. The important factor is the leader buys into it, hook, line, and sinker.

Selling the vision

The next step, which in fact never stops, is to constantly sell the vision. This takes energy and commitment, as few people will immediately buy into a radical vision, and some will join the show much more slowly than others. The transformational leader thus takes every opportunity and will use whatever works to convince others to climb on board the bandwagon.

In order to create followers, the transformational leader has to be very careful in creating trust and their personal integrity is a critical part of the package that they are selling. In effect, they are selling themselves as well as the vision.

Finding the way forward

In parallel with the selling activity is seeking the way forward. Some trans-formational leaders know the way, and simply want others to follow them. Others do not have a ready strategy, but will happily lead the exploration of possible routes to the Promised Land.

The route forward may not be obvious and may not be plotted in detail, but with a clear vision, the direction will always be known. Thus, finding the way forward can be an ongoing process of course correction and the transformational leader will accept that there will be failures and blind canyons along the way. As long as they feel progress is being made, they will be happy.

Leading the charge

The final stage is to remain up-front and central during the action. transformational leaders are always visible and will stand up to be counted rather than hide behind their troops. They show by their attitudes and actions how everyone else should behave. They also make continued efforts to motivate and rally their followers, constantly doing the rounds, listening, soothing, and enthusing.

It is their unswerving commitment as much as anything else that keeps people going, particularly through the darker times when some may question whether the vision can ever be achieved. If the people do not believe that they can succeed, then their efforts will flag. The transformational leader seeks to infect and reinfect their followers with a high level of commitment to the vision.

One of the methods the transformational leader uses to sustain motivation is in the use of ceremonies, rituals, and other cultural symbolism. Small changes get bog hurrahs, pumping up their significance as indicators of real progress.

Overall, they balance their attention between action that creates progress and the mental state of their followers. Perhaps more than other approaches, they are peopleoriented and behave that success comes first and last through deep and sustained commitment.

DISCUSSION

While the transformational leader seeks overly to transform the organization, there is also a tacit promise to followers that they also will be transformed in some way, perhaps to be more like this amazing leader. In some respects, then, the followers are the product of the transformation.

Transformational leaders are often charismatic, but are not as narcissistic as pure charismatic leaders, who succeed through a belief in themselves rather than a belief in others.

One of the traps of transformational leadership is that passion and confidence can easily be mistaken for truth and reality. While it is true that great things have been achieved through enthusiastic leadership, it is also true that many passionate people have led the charge right over the cliff and into a bottomless chasm. Just because someone believes they are right, it does not mean they are right.

Paradoxically, the energy that gets people going can also cause them to give up. Transformational leaders often have large amounts of enthusiasm which, if relentlessly applied, can wear out their followers. They also tend to see the big picture, but not the details, where die devil often lurks. If they do not have people to take care of this level of information, then they are usually doomed to fail.

Finally, transformational leaders, by definition, seek to transform. When the organization does not need transforming and people are happy as they are, then such a leader will be frustrated. Like wartime leaders, however, given the right situation they come into their own and can be personally responsible for saving entire companies.

2.9.3.6   AUTHENTIC LEADERSHIP

Although there are number of newly emerging theories, such as servant leadership, political leadership, contextual leadership, e-leadership, primal leadership, relational leadership, positive leadership, shared leadership, and responsible leadership. Authenticity has its roots in ancient Greek philosophy and descriptive words include transparent, reliable, trustworthy, real, and veritable. Positive psychologists refer to authenticity as both owning one's personal experiences (thoughts, emotions, or beliefs, the real me inside) and acting in accord with the true self (behaving and expressing what you really think and believe). Authentic leadership is defined as a process that draws from both positive psychological capacities and a highly developed organizational context, which results in both greater self-awareness and self-regulated positive behaviors on the part of leaders and associates, fostering positive self-development. The authentic leader is confident, hopeful, optimistic, resilient, transparent, moral, and future oriented, and gives priority to developing associates to be leaders.

2.10   TEAM IN ORGANIZATION

Teamwork is a way of life in the postmodern organization. Early discussions of the concept came from post-World War II Japanese management approaches and led to greater academic scrutiny in the human relations movement before being embraced by major US corporations. Through the years, many studies have documented the importance of teams for achieving organizational success. The basic premise of team work is that teams offer the best opportunity for better corporate performance in the form of increased productivity and profits. In other words, the synergistic benefits of teamwork are such that members of a team working coopera-tively with one another can achieve more than working independently. Large number of companies whether large or small, face challenges from a complex and global dynamic economy–challenges that have put in ques-tion the effectiveness of traditional methods. Some of the challenges are growing demands from customers for better quality, products, and services at lower prices; globalization, technological advances, and pressure from competitors and suppliers. According to some estimates, over 50% of all organizations and 80% of organizations with more than 100 employees use some form of teams. The thinking behind the team approach is that teams form the basic unit of empowerment–large enough for the collec-tive strength and synergy of diverse talents and small enough for effective participation and bonding.

2.10.1   SELF-MANAGED TEAMS

The challenge to succeed in a global economy has reached new levels, as companies are becoming more service oriented and under influence of changing environment and sustain competitive advantages. To meet these challenges, one structural approach that has been gaining ground is the self-managed work team (SMT). There is a general perception that these characteristics make self-managed teams more adaptive and proactive in their behavior than the traditional team. Self-managed teams have been used most often for manufacturing work, but they are finding increasing application in the service sector as well. Leadership skills for effective team meetings include being wellprepared for planning meetings, clear objectives, agenda, etc.

2.11   THE FOUR ROLES OF LEADERSHIP

The smooth current of business is history. Today, turbulence reigns in what Stephen R. Covey terms the "permanent white water world." The four roles of leadership deliver the tools, processes, and context to lead successfully–even in a time of turbulent change.

Participants will become more effective leaders by focusing their energies on the following four roles:

2.11.1 PATHFINDING: CREATING THE BLUEPRINT

Great leadership begins with clarity of thought and purpose. Stephen R. Covey says that all things are created twice–that the "mental creation precedes the physical creation." You would not build a home without a blue print. Similarly, it is folly to rush into action without understanding your purpose. The Path finding role helps you create a blueprint of action and ensure that your plans have integritybefore you act.

2.11.2 ALIGNING: CREATING A TECHNICALLY ELEGANT SYSTEM OF WORK

If pathfinding identifies a path, aligning paves it. Organizations are aligned to get the results they get. Think about that. If you are not getting the results you want, it is due to a misalignment somewhere, and no pushing, pulling, demanding, or insisting will change a misalignment. Therefore, as a leader, you must work to change your systems, processes, and structure to align them with the desired results you identified through pathfinding.

2.11.3 EMPOWERING: RELEASING THE TALENT, ENERGY, AND CONTRIBUTION OF PEOPLE

"Empowerment"is an overused term but underutilized in practice. Empowering isnot abandoning people, letting them "figure it out," on their own. Nor is it allowing individuals minute freedoms while controlling other aspects. True empowerment yields high trust, productive communication between individuals and teams, and innovative results where each member of the team feels welcome to bring his or her genius to the table.

2.11.4 MODELING: BUILDING TRUST WITH OTHERS–THE HEART OF EFFECTIVE LEADERSHIP

The four roles of leadership do not just teach you what a leader does, but who a leader is. You learn the essential balance between character and competence: an individual of high abilities will never be a true leader if his or her character is questionable. The processes and tools in the four roles of leadership enable you to get the results your organization needs while you model principles of effectiveness.

2.12 THE LEADERSHIP CHALLENGE

James Kouzes and Barry Posner developed a survey (The Leadership Practices Inventory) that asked people which, of a list of common characteristics of leaders, were, in their experiences of being led by others, the seven top things they look for, admire, and would willingly follow.

The results of the study showed that people preferred the following characteristics, in order:

Honest, forward-looking, competent, inspiring, intelligent, fair-minded, broad-minded, supportive, straightforward, dependable, cooperative, determined, imaginative, ambitious, courageous, caring, mature, loyal, self-controlled, and independent.

Kouzes and Posner identify these five actions as being the key for successful leadership:

MODEL THE WAY

Modeling means going first, living the behaviors you want others to adopt. This is leading from the front. People will not believe what they hear and what leaders say, but what they see leaders consistently do.

INSPIRE A SHARED VISION

People are motivated most not by fear or reward, but by ideas that capture their imagination.

Note that this is not so much about having a vision, but communicating it so effectively that others take it as their own.

CHALLENGE THE PROCESS

Leaders thrive on and learn from adversity and difficult situations. They are early adopters of innovation.

ENABLE OTHERS TO ACT

Encouragement and exhortation is not enough. People must feel and be able to act and then must have the ability to put their ideas into action.

ENCOURAGE THE HEART

People act best of all when they are passionate about what they are doing. Leaders unleash the enthusiasm of their followers with stories and passions of their own.

This overall process identified is clearly transformational in style, which again has a strong focus on followers. If something is to be leaked there must be a reward system setup to insure it.

2.13   GUIDELINES FOR LEADING MEETINGS

- Inform people about necessary preparations for a meeting: To ensure that people plan to attend the meeting, if they are wellinformed in advance a problem solving meeting can be made more effective.
- Describe the problem without implying the cause or solution: Problem can be defined objectively which does not assign blame for it to group members.
- Share essential information with group members: The leader should be careful to present facts and the amount of information that should be presented depends on the nature of the problem and the group's prior information.
- Give time for idea generation and evaluation: As decision might be important but due to time constraint, hasty decision can be taken, so leaders should plan meetings so that enough time is available for decision-making.

- Encourage and facilitate participation: Due to individual's personality differences, some member might be silent than others being more vocal to present their ideas and views.
- Encourage members to look for an integrative solution: If group is sharply divided in support of competing alternatives, leaders should develop ways to bring out an integrative solution that comprises the best features of the rival solutions.

2.14 ACTIVITIES OF SUCCESSFUL AND EFFECTIVE LEADERS IN TERMS OF

BE KNOW DO

- ➢ BE a professional. Examples: Be loyal to the organization, perform selfless service, and take personal responsibility. We must become the change we want to see–Mahatma Gandhi
- ➢ BE a professional who possess good character traits. Examples: Honesty, competence, candor, commitment, integrity, courage, straightforwardness, and imagination
- ➢ KNOW the four factors of leadership:Follower, leader, communication, and situation
- ➢ KNOW yourself and seek for self-improvement. Examples: Strengths and weakness of your character, knowledge, and skills
- ➢ KNOW human nature. Examples: Human needs, emotions, and how people respond to stress
- ➢ KNOW your job. Examples: Be proficient and be able to train others in their tasks
- ➢ KNOW and use full capacities of your organization. Examples: Where to go for help, its climate and culture, who the unofficial leaders are
- ➢ DO provide direction. Examples: Goal setting, problem solving, decision-making, and plaiting
- ➢ DO implement. Examples: Communicating, coordinating, supervising, and evaluating
- ➢ DO motivate. Examples: Develop moral and esprit in the organization, train, and coach
- ➢ Be technically proficient: As a leader, you must know your job and have a solid familiarity with your employees' tasks

> Seek responsibility and take responsibility for your actions: Search for ways to guide your organization to new heights. And when things go wrong, they always do sooner or later–do not blame others. Analyze the situation, take corrective action, and move on to the next challenge
> Make sound and timely decisions:Use good problem-solving, decision-making, and planning tools
> Know your people and look out for their well-being:Know human nature and the importance of sincerely caring for your workers
> Keep your workers informed:Know how to communicate with not only them, but also seniors and other key people
> Develop a sense of responsibility in your workers:Help to develop good character traits that will help them carry out their professional responsibilities
> Ensure that tasks are understood, supervised, and accomplished: Communication is the key to this responsibility.
> Train as a team: Although many so called leaders call their organization, department, section, etc. a team, they are not really teams... they are just a group of people doing their jobs
> Environment: Every organization has a canicular work environment, which dictates to a considerable degree how its leaders respond to problems and opportunities. This is brought about by its heritage of past leaders and its present leaders.

2.15 GOALS, VALUES, AND CONCEPTS

> Leaders exert influence on the environment via three types of actions:
> The goals and performance standards they establish.
> The values they establish for the organization.
> The business and people concepts they establish.

Successful organizations have leaders who set high standards and goals across the entire spectrum, such as strategies, market leadership, plans, meetings and presentations, productivity, quality, and reliability.

Values reflect the concern the organization has for its employees, customers, investors, vendors, and surrounding community. These values define the manner in how business will be conducted.

Concepts define what products or services the organization will offer and the methods and processes for conducting business.

These goals, values, and concepts make up the organization's "personality" or how the organization is observed by both outsiders and insiders. This personality defines the roles, relationships, rewards, and rites that take place.

2.16   ROLES AND RELATIONSHIPS

Roles are the positions that are defined by a set of expectations about behavior of any job incumbent. Each role has a set of tasks and responsibilities that may or may not be spelled out. Roles have a powerful effect on behavior for several reasons, to include money being paid for the performance of the role, there is prestige attached to a role, and sense of accomplishment or challenge.

Relationships are determined by a roles tasks. While some tasks are performed alone, most are carried out in relationship with others. The tasks will determine who the role-holder is required to interact with, how often, and toward what end. Also, normally the greater the interaction, the greater the liking, this in turn leads to more frequent interaction. In human behavior, it is hard to like someone whom we have no contact with, and we tend to seek out those we like. People tend to do what they are rewarded for, and friendship is a powerful reward. Many tasks and behaviors that are associated with a role are brought about by these relationships. That is, new task and behaviors are expected of the present role holder because a strong relationship was developed in the past, either by that role holder or a prior role holder.

2.17   DETERMINATION OF EFFECTIVE LEADERSHIP: VARIOUS DIMENSIONS

Although the above noted theories have more than proved the various aspects on the different traits and behaviors with respect to leadership styles of equally different individuals. There have been other attempts such as those focusing on different dimensions of leadership, which too need equal emphasis.

Thus, the researchers have found two such dimensions, the first is the dimension of "initiating structures" which takes into account the concern for organizational tasks, while the second dimension of consideration, takes into account concern for respective individual and their interpersonal relations. A brief into each of these dimensions reveals that die "initiating structures" includes within it activities, such as planning, organizing, and defining the various tasks and work assigned to the respective individuals, for example how a particular work is accomplished in an organization. On the other hand "consideration" takes into account the emotional requirements of the individual, their social requirements of the satisfaction and self-esteem which influences the performance of the respective individual.

In similar context, the dimensions of consideration and initiating structures have been differently defined by equally different researchers, and given different names. For example, some researches categorize this and name these dimensions as effectiveness and efficiency. In similar context, other researches note the same characteristics as goal achievement and group maintenance. On addition, other writings note them as instrumental and expressive needs, while others still put the characteristics of initiating structures and consideration as system- or person-oriented behaviors. Perhaps the assessment instrument developed, namely the "Leader Behavior Description Questionnaire (LBDQ)," has been the most widely used. As this finding alone noted that "effective leadership behavior tends most often to be associated with high performance on both dimensions," clearly implying that effective leaders were able to handle and cope both the given tasks as well as the various human aspects of the organization.

Hence, the above findings and researches on effective leadership and their history have differentiated between leaders and followers. These studies also reveal the differences between effective and noneffective leaders, which in turn led to the various dimensions, such as "initiating structures and consideration," as also explained in the preceding paragraphs (English, 2004).

Incomplete human beings become defective managers. Surviving in an ethics-less corporate culture requires executives to pursue their full potential. The first step is to embrace ethics through effective and appropriate leadership actions and behaviors.

The role of leadership: Any organization can establish a code of ethics; however, without the proper tone at the top (leadership), a compliance program will fail.

An example of a tool to enable the sustainability of such a compliance program is presented in the Exhibit: Leadership Diamond Realisms, as developed by Peter Koestenbaum. A sustainable compliance program internalizes ethical behavior and the proper role of leadership within the organization. Without that internalization, Sarbanes-Oxley becomes a nuisance to those unethical minds that will instinctively look for ways around it. Sustainable compliance programs like Leadership Diamond Realisms focus on key issues that are fundamental to any leader: freedom, principle, realism, grand strategy, and accountability.

2.17.1 FREEDOM

The foundation of successful leadership is complete understanding and application of the fact that human beings have free will. Free will makes ethics possible. It is the source of our power and the origin of our anxiety. Leadership coaching helps leaders convince personnel to think, feel, and behave in ways that help the goal. Leadership involves learning and teaching the ascending ladder of freedom, free will, consequences, responsibility, ownership, and accountability.

2.17.2 LEADERS CHOOSE PRINCIPLE

Leaders choose to live by Prince le. The 18th-century German ethicist Immanuel Kant wrote, two things fill the mind with ever-increasing wonder and awe, the more often and the more intensely the mind of thought is drawn to them: "The starry heavens above me and the moral law within me." Authentic leaders reject greed and selfishness, narcissism and naive values, and embrace the things that matter most: What is enduring, genuinely worthy, honest and generous, and what feels clean.

2.17.3 REALISM

It is a way of life. Being fully in touch with the real world is one definition of mental health. Realism is more than the numbers. It means never lying to oneself or denying the truth about oneself, as threatening as that is.

2.17.4 GRAND STRATEGY

One mark of an authentic leader is the commitment to a grand strategy. An exercise that develops this insight is to consider a major news story and ask: What deep lessons does this have for you in how you conduct your business and your life? What messages can you derive from an enlarged perspective of this or other monumental and historic events?

2.17.5 ACCOUNTABILITY

Being a true leader requires taking accountability for ones decisions and actions. What follows from setting an example in both word and deed is holding other people accountable for their free will-based decisions and actions.

2.18 HOW FOUNDERS/LEADERS EMBED AND TRANSMIT CULTURE

2.18.1 CULTURE FORMALLY DEFINED

A pattern of shared basic assumptions that the group learned as it solved its problems of external adaptation and internal integration that has worked well-enough to be considered valid and, therefore, to be taught to new members as the correct way you perceive, think, and feel in relation to those problems.

- The problem of socialization: teaching newcomers
- The problem of behavior
- Can a large organization have one culture? or subcultures?

2.18.2 CULTURE BEGINNINGS AND THE IMPACT OF FOUNDERS AS LEADERS

- Spring from three sources:

   1. Beliefs, values, and assumptions of founders
   2. Learning experiences of group members
   3. New beliefs, values, and assumptions brought by new members

2.18.2   IMPACT OF FOUNDER–MOST IMPORTANT

- Organizations do not form spontaneously or accidently.
- Leader assumptions are "taught" to the group.
- Things tried out are leader-imposed teaching.

2.18.2   THE PROCESS OF CULTURE FORMATION IS THE PROCESS OF CREATING A SMALL GROUP

1. Single person (founder) has idea
2. Founder brings in one or more people and creates core group. They share vision and believe in the risk
3. Founding group acts in conmen, raises money, work space
4. Others are brought in and a history is begun

   ➤ Jones an example of "visible management"

Cultures do not survive if the main culture carriers depart and if bulk of members leave.

- ➤ Smithfield staged things and then left them to the members.
- ➤ Murphy of Action: Total consensus had to be met. Open office landscape.

2.18.3   HOW DO LEADERS GET THEIR IDEAS IMPLEMENTED?

- Socialization
- Charisma
- Acting, by doing, exuding confidence

2.18.4   WHAT LEADERS PAY ATTENTION TO, MEASURE, AND CONTROL

- ➤ What leader systematically pays attention to communicates major beliefs.

  1. What is noticed
  2. Comments made

3. Casual questions and remarks
4. Becomes powerful if leader sees it and is consistent

➤ If leader is unaware and inconsistent then confusion can ensue
➤ Consistency is more important than intensity of attention
➤ Attention is focused in part by the kinds of questions that leaders ask and how they set the agendas for meetings
➤ Emotional reactions
➤ Important what they do not react to

2.18.5   LEADER REACTIONS TO CRITICAL INCIDENTS AND ORGANIZATIONAL CRISES

➤ In crisis: How do they deal with it?

1. Creates new norms, values, working procedures, reveals important underlying
2. Assumptions

➤ Crises are especially important in culture creation
➤ Crisis heightens anxiety, which motivates new learning
➤ A crisis is what is perceived to be a crisis, and what is defined by leader
➤ Crisis about leader, insubordination, tests leader

2.18.6   OBSERVED CRITERIA FOR RESOURCE ALLOCATION

➤ How budgets are created reveals leader assumption
➤ What is acceptable financial risk?
➤ How much of what is decided is all inclusive? Bottom up? Top down?

2.18.7   DELIBERATE ROLE MODELING, TEACHING, AND COACHING

➤ Own visible behavior has great value for communicating assumptions and values to others
➤ Video tape is good
➤ Informal messages are very powerful

2.18.8 OBSERVED CRITERIA FOR ALLOCATION OF REWARDS AND STATUS

> ➢ Members learn from their own experience with promotions, performance appraisals, and discussions with the boss.
> ➢ What is rewarded or punished is a message.

KEYWORDS

- leadership
- achievement
- power
- affiliation
- path–goal theory

CHAPTER 3

SELF-EFFICACY

CONTENTS

3.1 POSITIVE ORGANIZATIONAL BEHAVIOR

After paying attention to much traditional view of psychology in clinical and abnormal form, in the recent years, psychologists are focusing more on positivism and on new concepts of psychology and a new movement of psychology emerged in the form of more optimistic view, that is, positive psychology. The term positive organizational behavior or simply, POB defined by Luthans (2011), "the study and application of positive oriented human resource strengths, and psychological capacities that can be measured, developed, and effectively managed for performance improvement in today's workplace."

3.2 POSITIVE PSYCHOLOGY

The positive psychology movement under the leadership of Martin Seligman came with an observation several years ago with a concern that we are giving more attention to negative aspects of behavior and not giving importance to positive feelings and strengths of behavior. Rather than focusing on pathological functioning the focus is on optimal human functioning by the positive psychologists.

The three levels of positive psychology identified by Seligman and Csikszentmihalyi are:

Value subjective experiences which includes the emotions of past in form of well-being, contentment, and satisfaction, for the future hope and optimism and for the present flow and happiness comes in these experiences.

Positive individual traits include courage, capacity for love and vocation, interpersonal skill, aesthetic sensibility, forgiveness, perseverance, future mindedness, originality, spirituality, wisdom, and high talent.

Civic virtues and the institutions include responsibility, civility, nurturance, moderation, tolerance, and work ethic.

These very positive emotions have an important goal and implication in our life not for the well-being, family life, social or educational life but also from organizational point of view. Researches have shown that there is correlation between health, relationships, and work. This could be understood by the fact that one's health (both physical and mental) affects relationships (both personal and professional) and relationships affects health, their health affects work (both satisfaction and performance), and work affects their health and so forth.

3.2.1 WORK PERFORMANCE FROM PSYCHOLOGICAL CAPITAL

- ✓ Relationships
- ✓ Social networks
- ✓ Friends/life partner
- ✓ Health: which includes

  - Physical
  - Mental

3.2.2 SELF-EFFICACY

We can describe the self-concept as how individuals view themselves as physical, social, or moral and spiritual beings. Self-concept of an individual is shaped up by many factors and influences on one's life, which includes culture, family norms, siblings etc. Sociologist Viktor Gecas defines self-concept as "the concept the individual has of himself as a physical, social, and spiritual or moral being."In other words, because you have a self-concept, you recognize yourself as a distinct human being. Two related and crucial aspects of the self-concept are self-esteem and self-efficacy. People high in self-esteem consider themselves as capable, acceptable, and worthwhile and have less doubts about themselves. Some OB research suggests that whereas high self-esteem can boost performance and satisfaction outcomes, when under pressure, people with high self-esteem may become boastful and act egotistically.

Self-efficacy sometimes called the "effectance motive" is a more specific form of self-esteem; it is an individual's belief about the likelihood of success completing a specific task. You could be high in self-esteem yet have a feeling of low self-efficacy about performing a certain task, such as public speaking.

Self-efficacy is the first and most theoretically developed and researched positive organizational behavior construct. In positive organizational behavior, we tend to use the term self-efficacy interchangeably with confidence.

3.2.3   THE THEORETICAL BACKGROUND AND MEANING OF EFFICACY

The concept of self-efficacy was originally given by Professor Albert Bandura at Stanford University: A set of ideas embedded within his more general social cognitive theory of personality (Bandura, 1997, 1999). In any domain of functioning, our beliefs ofefficacy vary in their level, strength, and generality. The social cognitive theory emphasizes the role of observational leaning, social experience, and reciprocal determinism in the development of personality.

With Bandura's model there is a reciprocal interaction between perceived self-efficacy, behavior, and the external environment. Predicting behavior in a given setting is best done with tailor made domain specific measures of self-efficacy, usually single item 100 point rating scales. Self-efficacy beliefs assessed in this way are far more powerful predictors of behavior than general measures of self-concept or self-esteem (Bandura, 1997).

Self-efficacy refers to an individual's belief that he or she is capable of performing a task (Gist, 1987). The greater a person's self-efficacy the more confidence he or she has to succeed in a task. People with low self-efficacy are more likely to reduce their effort or cease an activity, while those high in self-efficacy will try harder to meet the challenge in difficult circumstances (Gist & Mitchell, 1992). It is also said that those high in self-efficacy are better able to handle negative feedback(

The relationship between self-efficacy and performance is a cyclical one. Efficacy-performance cycles can spiral upward toward success or downward toward failure. Researchers have documented strong linkages between high self-efficacy expectations and success in widely varied physical and mental tasks, anxiety reduction, addiction control, pain tolerance, illness recovery, avoidance of seasickness in naval cadets, physical exercise, and stress avoidance. Although self-efficacy sounds like some sort of mental magic, it operates in a very straightforward manner.

3.3   BASIC SOCIAL LEARNING CONCEPTS

1. People can learn through observation.
2. Mental states are important to learning: Environment reinforcement is not the only factor to influence learning and behavior. The cognition or mental processes are also important for learning.

3. Learning does not necessarily lead to change in behavior: While behaviorists believed that learning led to a permanent change in behavior, observational learning demonstrates that people can learn new information without demonstrating new behaviors. Factors involving both the model and the learner can play a role in whether social learning is successful, the following steps are involved in modelling process:

- Attention
- Retention
- Reproduction
- Motivation

On-the-job research evidence encourages managers to nurture self-efficacy, both in themselves and in others. In fact, a meta-analysis encompassing 21,616 subjects found a significant positive correlation between self-efficacy and job performance. Self-efficacy can be boosted in the work place through careful hiring, challenging assignments, training and coaching, goal setting, supportive leadership and mentoring, and rewards for improvement.

Boeing's CEO, James McNerney, recently offered this perspective:

"Success and achievement can feed on themselves. It feels good to keep succeeding. It feels great to see the people you work with grow and achieve. Maybe the ignition happens when you're younger, and then it feeds on itself. The next question is how you give it to people who weren't fortunate enough to have it given to them when they were young. It gets back to leadership attributes-expect a lot, inspire people, ask them to the the values that are important to them at home or at church and bring them to work."(Colvin, 2006)

For effective training, self-efficacy can be used, according to Robert F. Mager in "No Self-Efficacy, No Performance" (1992) sources include:

1. Mastery experience and performance attainment: The first most effective way of developing a sense of self-efficacy is through mastery experiences. Performing a task successfully strengthens our sense of self-efficacy. However, failing to adequately deal with a task or challenge can undermine and weaken self-efficacy.
2. Vicarious experience and modeling: Witnessing other people successfully, completing a task is another important source of

self-efficacy. According to Bandura, "Seeing people similar to oneself succeed by sustained effort raises observers' belief that they too possess the capabilities master comparable activities to succeed."

3. Social persuasion: Employees can be persuaded to believe that have the skills and capabilities to succeed. Getting verbal encouragement from others help people overcome self-doubt and instead focus on giving their best effort to the task at hand.

4. Physical and psychological arousal: Our own responses and emotional reactions to situations also play an important role in self-efficacy. Moods, emotional states, physical reactions, and stress levels can all impact how a person feels about their personal abilities in a particular situation.

As self-efficacy is directly tied to performance, some of the common practices engaged from leaders for enhancing self-efficacy are:

- *Recruiting applicant with high self-efficacy*
  Selecting individuals who have higher self-efficacy beliefs will increase the probability that they get more motivated to perform well.
- *Ensuring job demands that are appropriate*
  Managers must see the job demands and must be appropriate as employees who gain mastery completing complex challenging jobs, increase self-efficacy, and when job gets boring it leads to a decrease in self-efficacy.
- *Improve training and professional developmental practices*
  Leaders can develop professional development opportunities for all employees.
- *Enhance self-management*
  Encourage employees to set realistic personal goals, delineate clear priorities, and enhance time management skills.
- *Set reasonable goals and expectations*
  Goals that are challenging and attainable lead to both the highest performance levels and more resilient self-efficacy beliefs. Break larger goals into tangible steps.
- *Improving coaching strategies*
  The quality and quantity of feedback, guidance, training, and education should be improved and accurate for the team.

- *Confidence in employees*
  A leader should communicate and foster confidence in employees in various ways to have the impact on performance.
- *Mentoring and improving leadership*
  Identify your top performers and promote them to leadership or mentoring positions.
  Encourage managers and team leaders to engage in supportive leadership activities.
- *Acknowledgement and Reward*
  Both large and small contributions should be acknowledged and rewarded. Today a large number of leaders are using it on different platforms for enhancing other employee's self-efficacy to perform similar tasks.

Increasing self-efficacy in both individuals and teams should be made a managerial priority. When executed properly, increasing employee self-efficacy will lead to enhanced capabilities of employees and will have a major impact on the performance of individual employees as well as team.

KEYWORDS

- **self-efficacy**
- **self-management**
- **goals**
- **expectations**
- **learning**

CHAPTER 4

TRUST

CONTENTS

4.1   WHAT IS TRUST?

"Trust is the lubrication that makes it possible for organizations to work"–
Warren Bennis

Ralph Waldo Emerson explained the importance of trust in his 1994 essay, prudence and emphasized that trust relationship can develop only after one has carefully assessed the "present times, persons, property, and existing forms" of organizations.

The trust in organizations can be understood in several ways. One form of trust is interorganizational trust, or the trust between two organizations. Intraorganizational trust, a term that is used in different ways: Some researchers focus on the relationship between workers and their immediate superiors (e.g., supervisors), while others look at the relationship between workers and those running the organization (e.g., senior leaders; Dirks & Ferrin, 2002). The role of interpersonal trust within work groups and work teams can also be seen as an aspect of organizational trust.

Three descriptions of trust are:

- "The belief in the integrity, character, and ability of a leader" (Robbins & Coutler, 1999)
- "Reciprocal faith in one's intentions and behaviors" (Kreitner & Kinicki, 1998)
- "A confidant reliance on the integrity, honesty, or justice of another" (Funk & Wagnalls, 1985)

4.2   TYPES OF TRUST

Trust may be classified as basic, simple, blind, or authentic.

- Basic trust is the ability and willingness to meet people without inordinate suspicion, the ability to talk comfortably to and deal with strangers, and the willingness to enter into close relationships.
- Simple trust is the utter absence of suspicion: It demands no reflection, no conscious choice, no scrutiny, and no justification.
- Blind trust has been exposed to violation and betrayal but refuses to believe it has occurred. Blind trust denies the possibility that anything could shake or betray the trust.

- Authentic trust is fully self-aware, cognizant of its own conditions and limitations, open to new and even unimagined possibilities, based on choice and responsibility rather than the mechanical operations of predictability, reliance, and rigid rule following. Authentic trust leads to productive organizational relationships. Authentic trust is well aware of the risks and willing to confront distrust and overcome it. This type of trust evolves over time, starting with small actions and progressing to full strength based on individual experiences.

  Reina and Reina explain the "capacity for trust" in terms of three elements:

  Competence trust

  Contractual trust

  Communication trust

Why do people trust? Some say human trust is instinctual and evolved from the willingness to share food in hunter-gatherer societies (Nooteboom 2003). Others argue that people trust out of fear of punishment for not doing what is expected of them; to achieve self-interests; to demonstrate empathy, identification, and friendship; or simply because they genuinely want to trust (Greenburg & Baron, 2003, Nooteboom, 2003 & Reynolds, 1997). A popular model of trust by Mayer, Davis, and Schoorman suggests that three major factors determine organizational trust: Characteristics of the trustor, characteristics of the trustee, and the perceived risk.

4.3 BARRIERS TO BUILDING A CULTURE FOR TRUST IN ORGANIZATIONS

Employees who are incompetent or perceive others incompetent (I am OK, you are not OK and I am not OK but you are OK)

- Organizations with inflexible or inconsistent policies and rules
- Volatile personalities of employees
- Different goals of employees in respect of organization's goals
- Strong leadership can help organizations overcome these barriers to organizational trust.

4.4 LEADERSHIP AND ORGANIZATIONAL TRUST

A leader is a key to develop trust in the organizations. As trust is a reciprocal process, leaders can establish it by practicing it in their decisions and behavior. Team building in organizations contributes to trust building as interdependence creates the dynamic for reciprocity. Leadership in organization is either direct or top leadership. Dirks and Ferrin stated that the issue of leadership reference (supervisory versus top leadership) is important because it may provide guidance on whether an organization should focus resources on establishing trust in supervisors or in its senior leadership (Dirks & Ferrin, 2002). In addition, McCarthy found that supervisors play an important role in facilitating senior organizational leadership: Supervisors act as a medium between top leadership and front-line employees.

Empirical evidence suggests that trust in supervisors is related to job performance, altruism, job satisfaction, interactional justice, procedural justice, and participative decision-making.

4.5 MEASURING ORGANIZATIONAL TRUST

Kouzes and Posner identified four questions to measure one's trust worthiness as leader.

Is my behavior erratic or predictable?

Do I communicate clearly or carelessly?

Do I treat promises seriously or lightly?

Am I forthright or dishonest?

Employees' trust for the organization can be assessed by following statements given by Gabarro and Athos:

1. I believe my employer has high integrity
2. I can expect my employer to treat me in a consistent and predictable fashion
3. In general, I believe my employer's motives and intentions are good
4. I think my employer treats me fairly
5. Managers from my organization are open and upfront with me

Other more detailed measures of organizational trust are also available. Most are based on a division of organizational trust divided into three parts:

- Affective trust, the perception of a positive relationship with the trustee
- Cognitive trust, the belief in the reliability and integrity of the trustee
- A willingness to be vulnerable (Dirks & Ferrin, 2002)

4.6   HOW TO DEVELOP TRUST IN ORGANIZATIONS?

A high trust culture is essential for adapting to continuous change and continuous improvement (Sendjaya & Sarros, 2002). For leaders following suggestions can be helpful in creating trust:

- Practice humane leadership. Make sure employees know you are aware of, sensitive to, and understanding of their individual feelings, thoughts and experiences. Assure them promises will be kept, confidences maintained, and sensitive information handled judiciously (Robbins & Coutler, 1999)
- Develop, communicate, and apply organizational vision and value statements to ensure compatible beliefs
- Be a role model. Followers will be motivated and develop trust for organizational commitments
- Apply policies, procedures, and rules consistently and equitably
- Implement open door policies and open book policies in organizational communication channels. Share the results of organizational assessments of work with employees to build a culture of openness
- Demonstrate faith in employees by reducing supervision and monitoring of employees, and by building organizational structures that encourage delegation of authority, responsibility, and team work
- Workshops can help employees to understand the different types of trust, learn how to build authentic trusting relationships
- Develop an organizational "collective identity" by having employees work together in the same or collocated buildings(Hartog, 2003)
- When problems are investigated, attempt to be made to what went wrong and why, rather than who was responsible

4.7   TRUST AND GLOBALIZATION

In today's global economy that is increasingly depended on virtual organizations, trust is essential because direct leaders often do not see their
employees (Handy, 1995). Unfortunately, most of the writing on virtual
organization and trust as well as culture and organizational trust is on the
theoretical level. La Porta, Lopez-de-Silanes, Shleifer, and Vishny found
evidence in cross section of countries that suggest the role of trust in organizational performance cuts across cultural lines (La Porta et.al, 1997). One
very essential dimension in examining these cultural differences is the individualism–collectivism continuum. Individualistic cultures like that of the
United States, focus on the self, while collectivist cultures like that of Japan,
focus on the group. It is assumed that organizational trust should be higher
in collectivist cultures. However, Yamagishi has found consistently this is
not the case. Huff and Kelly discovered that organizational trust and propensity to trust was higher in the United States than in Japan, Korea, Hong
Kong, Taiwan, China, and Malaysia. This might be due to the differences
in openness among these cultures. For example, Americans are three times
as likely to express discontent within the workplace as the Japanese are to
express such feelings (Galford & Drapeau 2002). For leaders it is essential to be aware about cultural differences and how to develop trust among
employees.

Acknowledgements/Additional Reading
Special thanks to Becky J. Starnes, Stephen A Truhon & Vikkie McCarthy
https://asq.org/hd/2010/06/a-primer-on-organizational-trust.pdf [Retrieved
on 23 February 2015].

KEYWORDS

- **organizational trust**
- **leadership**
- **decision-making**
- **globalization**
- **self-interest**

CHAPTER 5

WORK–LIFE BALANCE

CONTENTS

5.1 INTRODUCTION

Working 24×7 and including diversity in work force by fair sex entering workforce, and also with industries becoming more service oriented today, a new challenge for employees and organization behaviorist emerged as work–life balance, so time has become money now for many young executives to fulfill personal needs with professional commitments. It is found that not only women but men are also constantly juggling between different roles and in constant effort to be successful in every role drives to health issues and other psychological problems. The result of this constant striving is becoming one of the reasons for fast aging among today's generation. Several studies by sociologists are showing that even in developing countries like China and India, fertility rates are declining due to this imbalance in work and personal life as many youngsters are getting late marriages with pushing parenthood further and giving priority to professional life, creating imbalance in their lives. Many organizations are taking steps to help their employees to cater both needs by providing several work arrangements like job sharing, work from home, flexible work hours, and some are providing other recreational activities for their arrangements like gyms, cafeteria, better work conditions, sabbatical leaves, etc. to counter their constant stress. Needless to say, these solutions are just mere superficial remedies and a lot is required for the employees to boost up their moods and rectify their emotional distress which will lead to a healthy generation finally.

Work–life balance is a broad concept, including proper prioritization between "work" (career and ambition) on one hand and "life" (Health, pleasure, leisure, family, and spiritual development) on the other. Related, though broader, terms include "lifestyle balance" and "life balance."

5.2 WORK–LIFE BALANCE DEFINED

Despite the world wide quest for work–life balance today, very few have found an acceptable definition of the concept. Work–life balance does not mean an equal balance. Trying to schedule an equal number of hours for each of our various work and personal activities is unrewarding and unrealistic. Life is and should be more fluid than that.

The best individual work–life balancevaries over time, often on a daily basis. The right balance for today can probably be different than tomorrow.

For an individual who is married and has children, work–life balance is different than for a single or married individual without kids.

There is no perfect, one size fits all, balance we should be striving for. The best work–life balance is different for each of us as we all have different values, priorities, and different lives.

However, at the core of an effective work–life balance definition are two key concepts which are relevant for each of us. These are daily achievement and enjoyment, ideas almost deceptive in their simplicity. Most of us already have a good grasp on the meaning of achievement. But enjoyment does not mean laughing every time. It is pride, satisfaction, happiness, celebration, love, and a sense of well-being with all the joys of living. Trying to live a one-sided life makes so many so-called professionally "successful" people not happy as happy they should be (Bird, 2003).

The expression was first used in the late 1970s to explain the balance between an individual's work and professional life. In the US, this term was first used in 1986.

Many Americans are experiencing burnout due to overwork and increased stress. This condition is seen nearly in all occupations from blue collar workers to upper management. Over the past decade, rises in workplace violence, an increase in levels of absenteeism, as well as rising workers' compensation claims are all evidence of an unhealthy work–life balance.

Employee assistance professionals say there are many causes for this situation ranging from personal ambition and the pressure of family obligations to the accelerating pace of technology.

5.3 DUAL-CAREER FAMILIES

Balancing the demands of work and family can be an arduous task. Oakley (1974) estimated that mothers with young children work an average of 77 hours per week in the home. Adding a career to that workload can create extraordinary pressures on both women and men to have fulfilling work and home lives.

Silberstein (1992) concluded that most dual-career couples have a work-oriented lifestyle prior to the birth of children. However, once children are there, the dual-career family orientation undergoes profound deviated. The demand for family life increases suddenly and cannot be ignored.

Karambayya and Reilly (1992), for example, reported that more women than men accommodated their careers to fit their families. The felt need to make an accommodation influenced the women's choice of a career or a work site. Silberstein (1992) reported that the difference in the degree that wives and husbands accommodate their career for children has become a central marital tension. Dual-career couples often cite a lack of temporal control over their lives. There appears to be insufficient time to fulfill the obligations of both work and family. Dual-career couples often cite a lack of temporal control over their lives. There appears to be insufficient time for work and family. Silberstein (1992) reported that most marriages are aided by the self-fulfillment each spouse derives from the pursuit of career but likely to feel that at least personal relationship suffers.

5.4   WORK–LIFE BALANCE INITIATIVES

Work–life balance initiatives include flexible work arrangements (e.g., working from home, compressed work weeks, and flexible working hours), leave arrangements (e.g., maternity leave, paternity leave, and sabbatical leave), dependent care assistance (e.g., on-site daycare, subsidized day care, elder care, and referral to child care), and general services (e.g., employee assistant programs, seminars, and programs related to family needs)(Frone, 2003). Work–life balance initiatives give employees flexibility and help ensure that dependents are cared for while employees are at work. Both work-to-family conflict and family-to-work conflict can be reduced when employees use work–lifebalance initiatives (Thompson et al., 1999; Allen, 2001; Haar & Spell, 2001; Anderson et al., 2002).

5.5   DEMOGRAPHIC FACTORS AFFECTING THE USE OF WORK–LIFE BALANCE INITIATIVES

While consistent age differences in the overall number of work–life balance initiatives used have not been found, consistent patterns in the extent to which different initiatives are used at different ages have been identified. Career stage models suggest that younger employees are likely to have fewer external demands on their time as they have not established their families to the same extent as midlife employees and may not have the challenge of caring for aging dependents. Older employees have been found to make more use of dependent care support, such as childcare, paid maternity, and paternity leave, and eldercare than younger employees (Allen, 2000).

Younger employees have entered the workforce at a time when employ-ability is valued more than job security and may place a greater value on nonwork commitments or developing their careers through ongoing education (Finegold, Mohrman & Spreitzer, 2002). Younger employees have been found to make more use of initiatives such as flexitime, compressed work weeks, telecommuting, and working from home than older employees (Thompson et al., 1999; Allen, 2000).

5.6 ORGANIZATIONAL FACTORS AFFECTING THE USE OF WORK–LIFE BALANCE INITIATIVES

For employees to use work–life balance initiatives they must first be aware that those initiatives are offered by the organization. Awareness of initia-tive availability is likely to be associated with initiative use. Employees who are aware of the availability of more work–life balance initiatives will use more work–life balance initiatives. The availability of work–life balance initiatives does not always mean that these initiatives will be used. There may be unspoken rules, peer pressure or perceived negative conse-quences from the organization that inhibit employees from using available initiatives (Kirby & Krone, 2002). Thomson et al. (1999) investigated the effect of workplace culture on work family initiative use and found that managerial support, perceived career damage, and organizational time demands predicted the use of work family initiatives.

There is a need for further research into the organizational factors that influence employees' use of available work–life balance initiatives. Thompson et al. (1999) found that managerial support was the strongest predictor of work–lifebalance initiative use. Management can influence hours worked through the timing of meetings, deadlines, the scheduling of training and holidays, monitoring work, and role modeling long hours at work (Perlow, 1998). In contrast, family-supportive managers may provide staff with the flexibility to meet external commitments or may model good work–life balance.

5.7   STUDIES ON WORK–LIFE BALANCE

Russell and Cooper (1992) examined work–life balance using a sample of 153 employees in a large organization in New Zealand. Analysis of company policies identified sixteen work–life balance initiatives currently being offered. Employees were surveyed to determine the extent of their awareness and use of currently offered initiatives. Factors influencing work–life balance initiative use and employee outcomes for initiative use were investigated. Female employees and younger employees used more work–life balance initiatives while employees reporting higher levels of management support and supervisor support, and perceiving fewer career damage and time demands also used more work–life balance initiatives. No support was found for the role of coworker support on work–life balance initiative use. Initiative use was related to reduce work to family conflict.

Work-to-family conflict, family-to-work conflict, and commitment to the organization were related to intention to turnover. The results highlight the importance of workplace culture in enabling an environment that is supportive of work–life balance and consequently use of initiatives that are offered by the organization. The following results were seen after the conduction of research:

5.7.1   DEMOGRAPHIC CHANGES

Including the increase in the number of women in the workplace, dual-career families, single-parent families, and an aging population have generated an increasingly diverse workforce and a greater need of employees to balance work and home life (Russell & Cooper, 1992; Frone & Yardley, 1996; Hobson, Delunas & Kesic, 2001; Brough & Kelling, 2002). Conflict between work and home life has been linked to job dissatisfaction and turnover; and increasingly organizations are using work–life balance initiatives to recruit and retain key personnel. Employees may view work–life balance initiatives as enabling them to balance their work commitments with their nonwork commitments, while employers are likely to view these initiatives as key strategies that enable organizations to recruit and retain employees (Allen, 2001; Hill, Hawkins, Ferris & Weitzman, 2001; Anderson, Haar & spell, 2001; Coffey & Byerly, 2002; Haar, 2004).

But not all employees make use of the initiatives that are available to them even those initiatives would be helpful. The present research aimed to identify demographic and workplace factors that influence the extent to which employees use available work–life balance initiatives and whether the use of these initiatives impact work–life balance and other outcomes. Women tend to use more work–life balance initiatives than men (Thomson et al., 1999; Allen, 2001). Compared to employed fathers, employed mothers were more likely to use childcare, flexible working hours, job sharing, and the opportunity to work at home (Frond & Yardley, 1996; Department of Labour, 1999).

5.7.2 *EMPLOYEES WITH DEPENDENTS*

Employees with dependents have been found to have a greater need for work–life balance initiatives and to make more use of these initiatives than those without dependents (Thompson, Beauvais Lyness, 1999; Brough & Kelling, 2002). Frone and Yardley (1996) found that the age of the youngest child was significantly related to the importance of initiatives, such as flexitime, compressed workweeks, childcare, and working from home but not to reduced hours or job sharing, while the number of dependents was significantly related to the importance of childcare.

5.7.3 *MARRIED EMPLOYEES*

They are significantly more likely to use work–life balance initiatives than unmarried employees (Thompson et al., 1999; Allen, 2001). It is likely that in general employees who have partners will be greater users of work–life balance initiatives than employees who do not have partners.

5.7.4 *EMPLOYEES WITH LONGER TENURE*

They will use more work–life balance initiatives than employees with shorter tenure. Employees with longer service with an organization may be more likely to adjust their work commitments when nonwork commitments arise but may have greater responsibilities at work and be less able to take time off work to tend to nonwork demands.

Employees with longer tenures also tend to have greater nonwork demands (Kirchmeyer, 1992; Finegold et al., 2002). Employees with longer service may be more aware of available work–life balance and make more use of these initiatives.

5.7.5 ORGANIZATIONAL FACTORS AFFECT

The use of work–life balance initiatives is useful for employees who are aware of the availability of more work–life balance initiatives. For employees to use work–life balance initiatives they must first be aware that those initiatives are offered by the organization. Awareness of initiative availability is likely to be associated with initiative use. The availability of work–life balance initiatives does not always mean that these initiatives will be used. There may be unspoken rules, peer pressure or perceived negative consequences from the organization that inhibit employees from using available initiatives (Kirby & Krone, 2002).

Thompson et al. (1999) investigated the effect of workplace culture on work family initiative use and found that managerial support, perceived career damage, and organizational time demands predicted the use of work-family initiatives. There is a need for further research into the organizational factors that influence employees use of available work–life balance initiatives.

Thompson et al. (1999) found that managerial support was the strongest predictor of work–lifebalance initiative use. Management can influence hours worked through the timing of meetings, deadlines, the scheduling of training and holidays, and role modeling long hours at work (Perlow, 1998). In contrast, family supportive managers may provide staff with the flexibility to meet external commitments or may model good work–life balance.

5.8 WOMEN AND FAMILY

Gender differences regarding work–life balance

According to Sylvia Hewlett, President of the Centre for Work-Life Policy, if a woman takes time off to care for children or an older parent, employers tend to "see these people as less than fully committed. It's as though their identity is transformed."

Brett Graff, *Nightly Business Report* correspondent states that (because a woman may have trouble reentering the market or, if she does find a position, it will likely to be a lower position with less pay). "If you thought choosing a baby name was hard, you have yet to wrestle with the idea of leaving your career to be a full time mom or take care of an older parent…. Most will want to reenter, but will do so accepting low positions or lower wages." This circumstance only increases the work–life balance stress experienced by many women employees.

Research conducted by the Kenexa Research Institute (KRI), a division of Kenexa, evaluated how male and female workers perceive work–life balance and found that women are more positive than men in how they perceive their company's efforts to help them balance work and life responsibilities. The report is based on the analysis of data drawn from a representative sample of 10,000 US workers who were surveyed through Work Trends, KRI's annual survey of worker opinions.

The results indicated a shift in women's perceptions about work–life balance. In the past, women often found it more difficult to maintain balance due to the competing pressures at work and demands at home.

Work–life balance concerns of men and women alike similar discrimination is experienced by men who take time off or reduce working hours for taking care of the family. For many employees today–both male and female–their lives are becoming more consumed with a host of family and other personal responsibilities and interests. Therefore, in an effort to retain employees, it is increasingly important for organizations to recognize this balance.

5.8.1 WORK–LIFE BALANCE ISSUES AND THEIR INFLUENCE ON CHILDREN

An increasing number of young children are being raised by a childcare provider or another person other than a parent; older children are more likely today to come home to an empty house and spend time with video games, television, and the internet with less guidance to offset or control the messages coming from these sources.

No one knows how many kids are home after school without an adult, but they know the number is in millions. Also, according to a study by the National Institute of Child Health and Human Development, the "more time that children spent in child care, the more likely their sixth-grade teachers were to report problem behavior." The findings are the results of the largest study of child care and development conducted in the United States; the analysis tracked 1,364 children from birth. Responsibility of the employer companies have begun to realize how important the work–life balance is to productivity and creativity of their employees.

Employers can offer a range of different programs and initiatives, such as flexible working arrangements in the form of part time, casual, and telecommuting work. More proactive employers can provide compulsory leave, strict maximum hours, and foster an environment that encourages employees not to continue working after hours.

It is generally only highly skilled workers that can enjoy such benefits as written in their contracts, although many professional fields would not go so far as to discourage workaholic behavior. Unskilled workers will almost always have to rely on bare minimum legal requirements. The legal requirements are low in many countries, in particular, the United States. In contrast, the European Union has gone quite far in assuring a legal work–life balance framework, for example, pertaining to parental leave and the nondiscrimination of part-time workers.

According to Stewart Freidman a "one size fits all" mentality in human resources management often perpetuates frustration among employees. It is not an uncommon problem in many HR areas where, for the sake of equality, there is a standard policy that is implemented in a way that is universally applicable even though everyone's life is different and every individual needs different things in terms of how to integrate the different pieces. It is got to be customized.

Management can influence hours worked through the timing of meetings, deadlines, the scheduling of training and holidays, monitoring work, and role modeling long hours at work (Perlow, 1998). In contrast, family-supportive managers may provide staff with the flexibility to meet external commitments or may model good work–life balance.

Companies have begun to realize how important the work–life balance is to the productivity and creativity of their employees. Research by Kenexa Research Institute in 2007 shows that those employees who were more favorable toward their organization's efforts to support work–life balance also indicated a much lower intent to leave the organization, greater pride in their organization, a willingness to recommend it as a place to work, and overall higher job satisfaction.

KEYWORDS

- work–life balance
- dual-career families
- single-parent families
- organizational factors
- self-fulfillment

CHAPTER 6

GLOBALIZATION AND VIRTUAL WORK

CONTENTS

6.1 INTRODUCTION

Globalization is when an organization extends its activities to other parts of the world, actively participates in other markets, and competes against organizations located in other countries.

Businesses today are being buffeted by major force winds–globalization, liberalization, and technology among them. The end result is constant churning, competition, and change.

The Aditya Birla Group, seen today as India's first truly global corporation, has a significant presence in South East Asia, Africa, North America, Australia, and China. The group is present in Thailand, Indonesia, Malaysia, and Philippines in the Asian subcontinent, Egypt in Africa, Canada in North America, and has recently forayed into China and Australia.

Globalization influences several organizational behavior issuessome good, some not so good. It requires new structures and different forms of communication to assist the organization's global reach. Globalization simply refers to the process of international integration in various domains, including economic, cultural, and technological unification. Most organizations must achieve high performance within a complex and competitive global environment. Globalization refers to the complex economic networks of international competition, resource suppliers, and product markets.

Forces of globalization

> - Rapid growth in information technology and electronic communication
> - Movement of valuable skills and investments
> - Increasing cultural diversity
> - Implications of immigration
> - Increasing job migration among nations

Impact of multicultural workforce

> - Globalization is contributing to the emergence of regional economic alliances.
> - Important regional alliances:

   - European Union (EU)
   - North American Free Trade Agreement (NAFTA)
   - Asia-Pacific Economic Co-operation Forum (APEC)

> Outsourcing:

- Contracting out of work rather than accomplishing it with a full-time permanent workforce.

> Off shoring:

- Contracting out work to persons in other countries.

> Job migration:

- Movement of jobs from one location or country to another.

The advent of globalization presented a number of opportunities and challenges for organizations. Globalization can be perceived as a form of socioeconomic integration across several countries resulting in cross-border movement of human resources, capital, commodities, and ideas; which are all significant aspects in the business world (Gullen, 2008). The onset of globalization has entirely transformed various aspects of organizations in numerous domains, such as changes in structure, changes in the leadership strategy, and changes in the planning strategy. These changes stemmed from several factors associated with globalization, such as cross-cultural differences, surfacing of global managers, workplace diversity, and the creation of a level playing field (Mirjana, 2012). Of particular concern is the effect of globalization on organizational behavior, which primarily stems from cultural integration. Organization behavior is concerned mainly with the study of human behaviors in the context of an organization. This paper outlines the challenges that organizations do face on the organizational behavior front in the age of globalization.

The first challenge posed by globalization on organizational behavior is that globalization threatens the unification of corporate culture. With increasing globalization, it is becoming apparent that the organizational culture is increasingly becoming irrelevant, whereas the global culture is increasing its relevance; as a result, organizations are faced with the challenge of establishing a corporate culture that meets the demands of global culture. This imposes significant challenges several domains of organizations such as human resource strategies and increasing their sensitivity to diverse cultures (Parker, 2005). This is a significant challenge because organizations are compelled to align their organizational culture with global trends, and this threatens the unification of a corporate culture, which is a defining factor for organizational success. The challenge for organizations is striking a balance between the demands of corporate culture at the organizational level and global cultural trends. Therefore, it can be argued that globalization threatens the firm's fundamental values, norms, and disposition that shape the reasons for the firm's existence.

The next challenge that organizations face in the wake of globalization regarding organizational behavior relates to the challenges of managing diversity. It is apparent that globalization increases workplace diversity, which translates to several implications for the managers and the entire organization. In addition, it is apparent that globalization tends to create a divergence between the organizational culture and global culture, which results in significant challenges for firms. For instance, in the wake of globalization, organizations are faced with the challenge of analyzing organizational cultures in order to change them and facilitate coordination of activities; understanding the what and how the various levels of culture are likely to be influenced by globalization; knowing how organizational culture and national culture interact to determine employee behaviors and shape the management philosophy (Mirjana, 2012).

Another challenge of globalization as regards organizational behavior relates to organizational relations, which refers to how individuals in the organization treat and interact with each other. It is apparent that globalization has negative impact on the facets of organizational relationships, such as collaboration, supportiveness, involvement, and openness and trust; this increases the difficulty in the establishment of a shared corporate culture. Organizational success in the age of globalization requires firms to adopt a global culture, which increases the difficulties associated with the leveraging diversity, addressing the gaps in corporate culture and global culture, and instilling core organizational values to strategies aimed at adopting a global culture (Mirjana, 2012).

Globalization threatens the creation of a shared corporate culture; challenges associated with managing diversity; and negative impacts on the aspects of organizational relationships, such as coloration, openness, involvement, trust, and supportiveness. As a result, firms are faced with the challenge of leveraging their corporate culture against the prevalence of global culture.

Globalization improves the organization's competitive advantage and creates new career opportunities, and potentially brings in new knowledge. However, globalization leads to more mergers and greater competition, which produces continuous change, layoffs, and other stressful consequences among employees. Any company, whether international or not, has to look at the way globalization has impacted the way we work…, "warns an executive at the global health care company GlaxoSmithkline." And even if you're not global, there is a competitor or customer somewhere who is".

Globalization emphasizes the need to recognize the possibility of effective OB practices in different cultures.

It will be increasingly difficult to discuss ethnic differences as interracial marriages increase. Gender diversity is playing a crucial role in organization. In the past decade, women workforce increased from 20% to 50%. Occupation wise gender diversity is also observed in workplaces. For example, during the 1960's, women represented only 17% of accountant jobs. Today, over half of the accountants are women.

Age groups represent another primary dimension of workforce diversity. Generation X and Generation Y have different level of responsibility and commitment toward organization.

The developing countries development agency, UNDP–United Nations Development Program–values geographical diversity in workforce. It also encourages women to apply.

Diversity represents both opportunities and challenges in organizations. For many businesses, a diverse workforce is necessary to provide better customer service in the global marketplace. When two companies from different nations get merged, executives at the two firms should understand cross-cultural sensitivity and team building procedures to minimize cross-cultural conflict.

6.2   EMERGING TRENDS IN EMPLOYMENT RELATIONSHIPS

After more than 100 years of relative stability, workforce and employment relationships are being redefined. Replacing the implied guarantee of lifelong employment in return for loyalty is a "new deal" called employability. Employees are concentrating on variety of work activities and skill upgradation that will keep them employed. Employability is an employment relationship in which people are expected to continually develop their skills to remain employed.

6.2.1 CONTINGENT WORK

Any job in which the individual does not have an explicit or implicit contract for long-term employment or one in which the minimum hours of work can vary in a nonsystematic way.

An example of contingent works is various teaching and nonteaching jobs both government and private.

Another example is Team Lease, one of India's largest HR outsourcing services companies in the space of employee leasing.

Some of the individuals like independence and reliance on knowledge that contingent work demands. But many contingent workers would rather be employed in stable, well-payingjobs. Contingent work affects organizational loyalty, career dynamics, and other aspects of organizational behavior.

6.2.3 TELECOMMUTING

Telecommuting (T/C) is a management tool that enables employees to work effectively from an alternative location. It not only reduces or eliminates daily commuting, but also positively impacts the organization's bottom line. Whether employees work from home, a satellite office, or a tele work center, forward thinking managers have found that telecommuting provides a terrific way to blend high employee morale with increased productivity and efficiency. Because it also helps in keeping a check on real estate and other overhead costs, telecommuting is fast becoming the preferred choice among executives in their never ending search for a competitive edge.

- All round the world, tens of millions of people have altered their employment relationship through telecommuting by working from home, usually with a computer connection to the office. Technology has changed some employees so completely from the employer's physical workspace that clients and coworkers are oblivious to their true locations. For instance, callcenter operators in Kansas work at home during snowstorms without the client's knowing the difference.

However, telecommuting poses a number of challenges for organizations and employees. Supervision at workplace is replaced by self-leadership.

6.2.4   VIRTUAL TEAMS

Contemporary competitive demands have forced many organizations to increase levels of flexibility and adaptability in their operations. A growing number of such organizations have explored the virtual environment as one means of achieving increased responsiveness. In particular, the use of virtual teams appears to be on the increase. Virtual teams are a product of modern times. They take their name from virtual reality computer simulations, where "it is almost like the real thing". Thanks to evolving information technologies such as the Internet, e-mail, instant messaging, videoconferencing, group ware, and fax machines, you can be a member of a work team without really being there. Virtual teams convene electronically with members reporting in from different locations, different organizations, and even different time zones. Our working definition of a virtual tem is a physically dispersed task group that conducts its business primarily through modern information technology

Virtual teams are cross-functional groups that operate across space, time, and organizational boundaries with members who communicate mainly through electronic technologies.

For virtual teams to perform well, three enabling conditions need to be established:

- Shared understanding
- Integration
- Mutual trust

Benefits of virtual teams include reduction in travelling expenses, allows more people to be included in labor pool, decreases air pollution, and physical handicaps are not concern.

6.3   DIVERSITY ISSUES

Even in one's own country, one will find himself working with bosses, peers, and other employees born and raised in different cultures. For working effectively with people from different cultures, leaders need to understand how their culture, geography, and religion cultures have shaped them, and how to adapt their management styles to their differences. Communication styles and motivational needs differ culturally, as what motivates one may not motivate others from a different culture. Individualistic cultures derive satisfaction more by fulfilling need of achievement rather than by fulfilling need of belongingness or affiliation. Communication style may be straightforward and open, which others may find uncomfortable and threatening.

Leaders at global companies, such as McDonald's, Disney, and Coca Cola have come to realize that economic values are not universally transferrable. Leadership practices need to be modified to reflect the values of the different countries in which an organization operates. Organizational leaders should examine their workforce to determine whether target groups have been underutilized. If groups of employees are not proportionally represented in top management, managers should look for any hidden barriers to advancement. Besides the mere presence of diversity in international work settings, there are international differences in how diversity is managed. Many countries such as India require specific targets and quotas for achieving affirmative action goals, whereas the legal framework in the United States specifically forbids their use. Globalization has certainly revolutionized organizations.

KEYWORDS

- socio-economic integration
- virtual work
- globalization
- workplace diversity
- telecommuting

REFERENCES

Adair, J. (2003). The Inspirational Leader. London: Kogan-Page.

Allen, T. (2001). Family-supportive work environments: The role of organizational perceptions. Journal of Vocational Behavior, 58(3), 414–435.

Avolio, B.J., Kahai, S., and Dodge, G. (2000). E-leadership and its implications for theory, research and practice. Leadership Quarterly, 11, 615–670.

Bandura, A. (1997). Self-Efficacy: The Exercise of Control. New York: Freeman.

Bendix, R. (1956). Work and Authority in Industry: Ideologies of Management in the Course of Industrialization. New York: Wiley.

Bennis, W. (1989). Why leaders can't lead. Training and Development Journal, 43(4), 35–39.

Bird, J. (2003). Work Life Balance. http://www.worklifebalance.com/work-life-balance-defined.html.

Blake, R., and Mouton, J. (1964). The Managerial Grid: The Key to Leadership Excellence. Houston: Gulf Publishing Co.

Bork, D. (1986). Family Business, Risky Business: How to Make It Work. New York: AMACOM.

Brough, P., and Kelling, A. (2002). Women, work & well-being: The influence of work-family and family-work conflict. The New Zealand Journal of Psychology, 31(1), 29–39.

Chatman, J.A., and Jehn, K.A. (1994). Assessing the relationship between industry characteristics and organizational culture: How different can you be? Academy of Management Journal, 37, 522–533.

Child, J. (1977). Organization: A Guide for Managers and Administrators. New York: Harper & Row.

Colvin, G. (2006). How one CEO learned to fly. Fortune, Oct 30, p. 100.

Czeisler, C.A. (2006). Sleep deficit: The performance killer. Harvard Business Review, 53–59.

Dirks, K.T., and Ferrin, D.L. (2002). Trust in leadership: Meta-analytic findings and implications for research and practice. Journal of Applied Psychology, 87(4), 611–628.

Drath, W.H., and Palau, C.J. (1994). Making Common Sense: Leadership as Meaning-Making in a Community of Practice. Greensboro, NC: Center for Creative Leadership.

Duncan, R. (1979).What is the right organizational design? Organizational Dynamics, Winter, 59–80.

Emerson, R.W. (1944). The Essays of Ralph Waldo Emerson (The Illustrated Modern Library). New York Random House, Inc.

Fiedler, F. (1967). A Theory of Leadership Effectiveness. New York: McGraw-Hill.

Finegold, D., Mohrman, S., and Spreitzer, G.M. (2002) Age effects on the predictors of technical workers' commitment and willingness to turnover. Journal of Organizational Behavior, 23, 655–674.

Frone, M.R. (2003). Work-family balance. In J.C. Quick and L.E. Tetrick (Eds.), Handbook of Occupational Health Psychology. Washington, D.C.: American Psychological Association, pp. 143–162.

Frone, M.R., and Yardley, J.K. (1996). Workplace family-supportive programmes: Predictors of employed parents' importance ratings. Journal of Occupational and Organizational Psychology, 69(4), 351–356.

Frone, M.R., Russell, M., and Cooper, M.L. (1992). Antecedents and outcomes of work-family conflict: Testing a model of the work-family interface. Journal of Applied Psychology, 77(1), 65–78.

Funk & Wagnalls. (1985). Standard Desk Dictionary. New York: Harper & Row.

Galford, R.M., and Drapeau, A.S. (2002). The Trusted Leader: Bringing Out the Best in Your People and Your Company. New York: Free Press.

Gallant, M. (2013). The Business of Culture: How Culture Affects Management Around the World, Talent Space Blog. http://www.halogensoftware.com/blog/the-business-of-culture-how-culture-affects-management-around-the-world [Retrieved on 13 November 2014].

Garvin, D.A. (1993). Building learning organization. Harvard Business Review, 71, (4), 78–91.

Gist, M.E. (1987). Self-efficacy: Implications for organizational behavior and human resources management. Academy of Management Review, 12, 472–85.

Gist, M.E., and Mitchell, T.R. (1992). Self-efficacy: A theoretical analysis of its determinants and malleability. Academy of Management Review, 17, 183–211.

Gullen, M. (2008). The Limits of Convergence: Globalization and Organizational Change in Argentina, South Korea, and Spain. Princeton: Princeton University Press.

Gutman, H. (1975). Work, Culture and Society in Industrializing America. New York: Knopf.

Haar, J.M. (2004). Work-family conflict and turnover intention: Exploring the moderation effects of perceived work-family support. New Zealand Journal of Psychology, 33(1), 35–39.

Haar, J., and Spell, C. (2001). Examining work-family conflict within a New Zealand local government organization. The New Zealand Journal of Human Resources Management, 1, 1–21.

Hagberg, R., and Heifetz, J. (2000). Corporate Culture/Organization Culture: Understanding and Assessment (online document). www.hednet.com.

Handy, C. (1995). Trust and the virtual organization. Harvard Business Review, 73(3), 40–50.

Hebden, J.E. (1986). Adopting an organization's culture: The socialization of graduate trainees. Organizational Dynamics, Summer.

Hemphill, J.K., and Coons, A.E. (1957). Development of the leader behavior description questionnaire. In R.M. Stodgill and A.E. Coons (Eds.), Leader Behavior: Its Description and Measurement. Columbus, Ohio: Bureau of Business Research, Ohio State University, pp. 6–38.

Hill, E.J., Hawkins, A.J., Ferris, M., and Weitzman, M. (2001). Finding an extra day a week: The positive influence of perceived job flexibility on work and family life balance. Family Relations, 50(1), 49–55.

Hobson, C.J., Delunas, L., and Kesic, D. (2001). Compelling evidence of the need for corporate work/life balance initiatives: Results from a national survey of stressful life-events. Journal of Employment Counseling, 38, 38–44.

House, R.J. (1996). Path-goal theory of leadership. Lessons, legacy, and reformulated theory. Leadership Quarterly, 7, 323–352.

Huff, L., and Kelley, L. (2003). Levels of organizational trust in individualist and collectivist cultures: A seven-culture study. Organizational Science, 14, 81–90.

Jaffee, D. (2008). Conflict at work throughout the history of organization. In C.K.W. De Dreu and M.J. Gelfand (Eds.), The Psychology of Conflict and Conflict Management in Organizations. New York: Lawrence Erlbaum Associates.

Jones, G.R. (2003). Organizational Theory, Design and Change: Text and Cases. Upper Saddle River: Prentice Hall.

Jones, G.R., and George, J.M. (2003). Contemporary Management, 3rd ed. New York, USA: McGraw-Hill, Inc.

Katz, D., and Kahn, R.L. (1978). Social Psychology of Organizations, 2nd ed. New York: John Wiley.

Kenexa Research Institute, July, 2007 [Retrieved on 27 May 2008].

Kirby, E.L., and Krone, K.J. (2002). "The policy exists but you can't really use it": Communication and the structuration of work-family policies. Journal of Applied Communication Research, 30(1), 50–77.

Kirchmeyer, C. (1992). Perceptions of nonwork-to-work spillover: Challenging the common view of conflict-ridden domain relationships. Basic and Applied Social Psychology, 13(2), 231–249.

Kirkpatrick, S.A., and Locke, E.A. (1991). Leadership: Do traits matter? *The Academy of Management Executive*, 5(2), 48–60.

Kouzes, J.M., and Posner, B.Z. (1987). The Leadership Challenge. San Francisco: Jossey-Bass.

Kreitner, R., and Kinicki, A. (1998). Organizational Behaviour, 4th ed. Boston: Irwin McGraw-Hill.

La Porta, R., Lopez-de-Silanes, F., Shleifer, A., and Vishny, R.W. (1997). Trust in Large Organizations. A Working Paper. NBER Working Paper Series.

Likert, R. (1967). The Human Organization: Its Management and Value. New York: McGraw-Hill.

Mager, R.F. (1992). No self efficacy, no performance. Training, 29(4), 32–36.

Maier, N.R.F. (1963). Problem-Solving Discussions and Conferences: Leadership Methods and Skills. New York: McGraw-Hill.

March, J. G., and Simon, H.A. (1958). Organizations. New York: Wiley.

Markus, M.L., Manville, B., et al. (2000). What makes a virtual organization work? Sloan Management Review, 42(1),13–26.

Mayer, R.C., Davis, J.H., and Schoorman, E.D. (1995). An integrative model of organizational trust. Academy of Management Review, 20, 709–734.

McCarthy, V.M. (2006). Facilitative Leadership: A Study of Federal Agency Supervisors. Paper presented at the annual meeting of the Midwest Political Science Association, Chicago, IL. http://www.allacademic.com/meta/p137946_index.html [Retrieved on 26 June 2008].

McClelland, D.C., and R.E. Boyatzis. (1982). Leadership motive pattern and long-term success in management. Journal of Applied Psychology, 67, 731–743.

McKee-Ryan, F.M., Song, Z., Wanberg, C.R., and Kinicki, A.J. (2005). Psychological and physical well-being during unemployment: A meta-analytic study. Journal of Applied Psychology, 53–76.

McKelvey, W. (1982). The Evolution of Organizational Forming in Ancient Mesopotamia, Organizational Systematics. University of California Press, Los Angeles, CA, pp. 295–33.

McKenna, E. (2000). Organizational Structure and Design, Business Psychology and Organisational Behaviour: A Student's Handbook, 3rd ed. Philadelphia, PA: Psychology Press, pp. 421–429.

Mirjana, R.M. (2012). Impact of Globalization on Organizational Culture, Behavior and Gender Role. New York: IAP.

Montgomery, D. (1979). Workers' Control in America: Studies in the History of Work, Technology and Labor Struggles. New York: Cambridge University Press.

Muchinsky, P.M. (2000). Psychology Applied to Work. Singapore, 6th ed. Wardsworth Thomson Learning, pp. 319–322.

Nooteboom, B. (2003). The trust process. In B. Nooteboom and F. Six (Eds.), The Trust Process in Organizations: Empirical Studies of the Determinants and the Process of Trust Development. Edward Elgar Publishing, Inc.

Nutt, P.C. (1999). Surprising but true: Half the decisions in organizations fail. Academy of Management Executive, 13(4), 75–90.

Oakley, A. (1974). Housewife. London: Allen Lane.

Pang, L. (2001). Understanding virtual organizations. Information Systems Control Journal, 6, 42–47.

Parker, B. (2005). Introduction to Globalization and Business: Relationships and Responsibilities. New York: SAGE.

Perlow, L. (1998). Boundary control: The social ordering of work and family time in a high-tech corporation. Administrative Science Quarterly, 43(2), 328–58.

Perrow, C. (1970). Organizational Analysis: A Sociological View. Belmont Wadsworth Publishing.

Pollard, S. (1965). The Genesis of Modern Management. Cambridge: Harvard University Press.

Reina, D.S., and Reina, M. (2008). Trust and Betrayal in the Workplace: Building Effective Relationships in Your Organization. Sans Francisco: Berrett-Koehler Publishers.

Reynolds, L. (1997). The Trust Effect Creating the High Trust, High Performance Organization. London: Nicholas Bradley Publishers.

Richards, D., and Engle, S. (1986). After the vision: Suggestions to corporate visionaries and vision champions. In J.D. Adams (Ed.), Transforming Leadership. Englewood cliffs, NJ: Prentice Hall.

Robbins, S.P., and Coulter, M.K. (1999). Management, 6th ed. New Jersey: Prentice Hall.

Robbins, S.P., and Judge, T.A. (2009). Organisational Behaviour, 13th ed. NJ: Prentice Hall.

Sendjaya, S., and Sarros, J. (2002). Servant leadership: Its origin, development, and application in organizations. Journal of Leadership and Organizational Studies, 9, 57–64.

Smith, J., and Gardner, D. (2007). Factors affecting employee use of work life balance initiatives. New Zealand Journal of Psychology, 36(1). http://www.psychology.org.nz/wp-content/uploads/36-1_Smith-Gardner_pg3.pdf [Retrieved on 21 February 2014].

Sonnefeld, J. (1988). The Hero's Farewell. New York, NY: Oxford University Press.

Stodgill, R.M. (1974). Handbook of Leadership: A Survey of Theory and Research. New York: The Free Press.

Survey: U.S. Workplace Not Family Oriented, Forbes, Feb 1, 2007 [Retrieved on 17 February 2007].

Suttle, R. (2015). Types of Organizational Structure in Management. http://smallbusiness.chron.com/types-organizational-structure-management-2790.html [Retrieved on 2 January 2015].

Tannenbaum, A.S., and Schmitt, W.H. (1958). How to choose a leadership pattern. Harvard Business Review, 36, March–April, 95–101.

The Family and Medical Leave Act of 1993, U.S. Department of Labor, March 1, 2007.

Thompson, C., Beauvais, L., and Lyness, K. (1999). When work-family benefits are not enough: The influence of work-family culture on benefit utilization, organizational attachment and work-family conflict. Journal of Vocational Behavior, 54(3), 392–415.

Whaples, R. (Ed.) Hours of Work in U.S. History. EH. Net Encyclopedia, Aug 2001 [Retrieved on 3 April 2007].

When It's Just You After School, Kids Health. 2007. The Nemours Foundation, April 4, 2007.

Wide Variation in European Maternity Benefits, HRM Guide, Feb17, 2007.

Yamagishi, T., Jin, N., and Miller, A.S. (1998). In-group bias and culture of collectivism. Asian Journal of Social Psychology, 1, 315–328.

Yukl, G.A. (1971). Toward a behavioural theory of leadership. Organizational Behavior and Human Performance, 6, 414–440.

Yukl, G. (1989). Managerial leadership: A review of theory and research. Journal of Management, 15(2), 251–289.

Yukl, G.A. (1989). Leadership in Organizations. Englewood Cliffs, NJ: Prentice Hall.

Zipkin, A. (2000, May 31). The Wisdom of Thoughtfulness. New York Times, pp. C1–C10.

http://www.referenceforbusiness.com/management/Tr-Z/Virtual-Organizations.html#ixzz3IsPlyP8W [Retrieved on 12 November 2014].

http://ssf.f15ijp.com/wiki/index.php/Work%E2%80%93life_balance.

PART II
RESEARCH AND CRITICAL ANALYSIS

Research on the relation between leadership effectiveness, perceived organizational climate, interpersonal trust, and self-efficacy among virtual workers in different organizations

CHAPTER 7

INTRODUCTION

CONTENTS

7.1 INTRODUCTION

According to The Economist Intelligence Unit's Foresight 2020 research, there are five key trends for the next 15 years, out of which, the two most important are globalization and knowledge management. Both will impact the structure, functioning, and distribution of teams. Because of organizations becoming more global, geographically dispersed team (GDT) and multicultural work scenario will be emerging at fast pace. Work will especially get broken down into smaller units to be managed by teams of specialists or individuals linked by technology (Cisco Report, 2006).

Human societies across the globe have established progressively closer contacts over many centuries, but recently the pace has dramatically increased. As technology advanced and liberalization came in early nineties in India, not only did it change individual lifestyles to a large extent but also bought the world closer. Jet airplanes, cheap telephone service, email, computers, huge oceangoing vessels, instant capital flows, all have made the world more interdependent than ever. Multinational corporations manufacture products in many countries and sell to consumers around the world. Money, technology, and raw materials move ever more swiftly across national borders. Along with products and finances, ideas and cultures circulate more freely. As a result, laws, economies, and social movements are forming at the international level. This globalization bought big change in organizations also. Virtual working is one of them.

The virtual organization has long been a neat theory. Now organizations throughout the world are turning it into reality.

It is only when the tide goes out you can see who's swimming naked. And so, it is only when times get tough that you see whose business and management models are most robust. The global downturn that we faced last year has exposed the frailties of many organizations and confirmed the strength of others.

One model that looks increasingly robust in difficult times is the virtual model–organizations that will orchestrate the activities of many independent actors rather than owning, employing, and controlling lots of people. The major virtue of virtuality is that, if it is done right, it creates truly agile organizations well equipped to negotiate the crumbling markets of recession (Birkinshaw, 2010).

Worker productivity in the United States fell unexpectedly in the second quarter according to the Labor Department. At the same time, labor costs increased. Specifically, worker productivity declined at an annual rate of 0.9% in the second quarter after seeing large improvements in 2009. Labor costs rose 0.2% in the quarter, marking the first increase since spring 2009. This is a contrast to the height of the recession, when companies slashed their payroll and employee productivity rose fairly dramatically. When worker output rises, so do living standards because companies can pay larger salaries without raising the cost of goods. How can companies combat the worker productivity slide? Companies that work from office buildings can turn to virtual office space.

Virtual work can also help companies boost productivity by attracting the most qualified staff, no matter where they live. An employee in Los Angeles can work for a company in Manhattan using virtual office technologies that generate greater productivity by effectively extending the company's operating hours.

What's more, a virtual office can lead to higher employee retention rates, removing the cost and productivity losses associated with hiring and training new workers.

Although this strategy of virtual work does not help companies in retail or restaurant industries because employees have to work on-site, the headquarters of these service-oriented companies can still leverage virtual office technologies to achieve productivity gains among administrative and executive staff (Claire, 2010).

A prominent organizational psychologist describes the nature of work as follows:

Consider the new paradigm of work—anytime, anywhere, in real space, or in cyberspace. For many employers, the virtual workplace, in which employees operate remotely from each other and from managers, is a reality now, and all indications are that it will become even more prevalent in the future (Casio, 1998).

It is becoming commonplace for organizations to have large numbers of employees who work off-site, telecommuting from a home office, phoning from a car or airplane while travelling on business, or teleconferencing from a hotel room or vacation spot.

This dramatic shift in where and how we work is an effect of the information age. Many jobs can be performed anywhere within electronic reach of the home office or the actual workplace, thanks to e-mail, voicemail, pagers, cell phones, laptop computers, and personal data systems.

To function efficiently and productively, these virtual workplaces require three types of information access (Casio, 1998).

- Online material that can be downloaded and printed.
- Databases on customers and products and automated central files, which can be accessed from remote locations.
- A means of tracking employees and their work assignments at any time of day.

The downside of electronically connected virtual workplaces is that employees are often expected to work, or to be available, beyond the normal working hours of the organization. Some companies require their employees to carry phones or beepers at all times, keeping them effectively tethered to the office, with no way to escape the demands of their jobs. As a writer for the *New York Times* noted, "The 24/7 culture of nearly round the clock work is endemic to the wired economy."

While teams are not a new phenomenon, they currently are a popular way for organizations to provide a structure that places power in the hands of employees as well as management. Many contemporary organizations have created team-based work structures that are significantly different than the hierarchical and control-based organizations of the industrial era. However, advances in communication technologies have dramatically changed the nature of teamwork. Traditional collocated groups are being replaced with virtual teams, distributed across boundaries of time, space, and organizational structures.

The Industrial Age was characterized by hierarchical organizations that relied on management direction and organizational departmentalization to provide order and consistency. Rules and auditing processes were important means of control. Employees' roles and responsibilities tended to be specialized and information typically went to management rather than to employees. Hard work was encouraged more than a balance between work and home life. Conservative improvements tended to be the norm because organizational controls typically inhibited risk taking (Fisher and Fisher, 1998).

Unlike rational organizational structures of the past, teams rely on employee empowerment rather than management control and direction. Team organizations have created work structures that are more democratic and flexible with a common mission of sharing responsibility for results and decisions between management and workers. The ideal team is characterized by a global rather than departmental focus. Problems are controlled at the source rather than by a separate policy function. Information tends to go to employees and there is more of an emphasis on work and home life balance as opposed to long hours. Continuous improvement is highly valued. Instead of promoting employees with highly specialized skills, team-based operations focus on creating flexible, cross-trained, and multi-skilled team members. Self-managing teams are said to be the key to leaner and more flexible organizations capable of adjusting rapidly to changes in the environment and technology (Fisher and Fisher, 1998).

Virtual teams are the next logical step in the evolution of organizational structures (Lipnack and Stamps, 1999). Presently, people work across internal organizational boundaries such as specialized functions and departments as well as external organizational boundaries such as alliances with vendors, industry associations, and even competitors. Virtual teams explore a new type of boundary-crossing organization utilizing technology and information (Geisler, 2002).

Townsend and colleagues (1998, p. 17) defined virtual teams as "groups of geographically and/or organizationally dispersed coworkers that are assembled using a combination of telecommunications and information technologies to accomplish an organizational task." In fact, these teams are used to accomplish a variety of critical tasks. PricewaterhouseCoopers, which has 45,000 employees in 120 countries, uses virtual teams to bring employees from around the globe "together" for a week or two to prepare work for a particular client. Whirlpool Corporation used a virtual team composed of experts from the United States, Brazil, and Italy during a 2-year project aimed at developing a chlorofluorocarbon-free refrigerator (Geber, 1995). Virtual teams offer many benefits. They allow organizations to access the most qualified individuals for a particular job regardless of their location, enable organizations to respond faster to increased competition, and provide greater flexibility to individuals working from home or on the road. There is little doubt that virtual teams will play a key role in the design of organizations in the new millennium (Bell & Kozlowski, 2002).

Previous studies have investigated issues in virtual teams (Yoong 2001, Suchan et. al. 2001). A multi-cultural team is a team whose members have different cultural backgrounds, for instance because they are from different countries. In a global marketplace, more and more companies need international presence; therefore, the need for creating virtual teams exists. By dynamically allocating people to projects based on expertise rather than location, organizations can more easily assign the most qualified people to appropriate projects without concern for the expense and wasted productivity caused by extensive travel or frequent relocation (Goldman, 2000).

Virtual teams are "groups of geographically, organizationally and/or time dispersed workers brought together by information and telecommunication technologies to accomplish one or more organizational tasks" (Powell, Piccoli, & Blake, 2004), or as Grundy (2004) states, working virtual means *working together apart*. Organizational change employing virtual working enables industries to be globally competitive, provide flexible workspace, and just-in-time responses (Powell, Piccoli, & Blake, 2004; Egea, 2006).

7.1.1 THE VIRTUAL WORKPLACE

A **virtual team**–also known as geographically dispersed team–is a group of individuals who work across time, space, and organizational boundaries with links strengthened by webs of communication technology. They have complementary skills and are committed to a common purpose, have interdependent performance goals, and share an approach to work for which they hold themselves mutually accountable. GDTs allow organizations to hire and retain the best people regardless of location. Members of virtual teams communicate electronically, so they may never meet face to face. However, most teams will meet at some point in time. A virtual team does not always mean teleworker. Teleworkers are defined as individuals who work from home. Many virtual teams in today's organizations consist of employees both working at home and small groups in the office, but in different geographic locations.

7.1.2 WHO ARE THE MEMBERS OF VIRTUAL TEAMS?

- Members can either be stable or change on an ongoing basis.
- Members can be in the same company or from various companies.
- Members can live in the same community or in different countries.

7.1.3   STRATEGIES FOR VIRTUAL TEAMS

The following tips come from research into virtual teamwork.

- Hold an initial face-to-face startup
- Have periodic face-to-face meetings, especially to resolve conflict and maintain team cohesiveness
- Establish a clear code of conduct or set of norms and protocols for behavior
- Recognize and reward performance
- Use visuals in communications
- Recognize that most communications will be non-verbal–use caution in tone and language

7.1.4   TECHNOLOGY SUPPORTING VIRTUAL TEAMS

Virtual teams are supported by both hardware and software. General hardware requirements include telephones, PCs, modems or equivalent, and communication links such as the public switched network (telephone system) and local area networks. Software requirements include groupware products such as electronic mail, meeting facilitation software, and group time management systems.

One way to think about teams is that teams are a network organization–a set of nodes and links–wherein the nodes are of course the team members and the links are the communications channels or primarily face-to-face interaction. In virtual teams, the nodes are the same–team members–whereas the links are primarily virtual (electronic) and software is used to mediate the interactions. In simple terms, then

virtual teams = teams + electronic links + groupware

7.1.5   WHY VIRTUAL TEAMS?

- Best employees may be located anywhere in the world.
- Workers demand personal flexibility.
- Workers demand increasing technological sophistication.

- A flexible organization is more competitive and responsive to the marketplace.
- Workers tend to be more productive–less commuting and travel time.
- The increasing globalization of trade and corporate activity.
- The global workday is 24 versus 8 h.
- The emergence of environments which require interorganizational cooperation as well as competition.
- Changes in workers' expectations of organizational participation.
- A continued shift from production to service/knowledge work environments.
- Increasing horizontal organization structures characterized by structurally and geographically distributed human resources.

7.1.6 BENEFITS OF VIRTUAL TEAMS

Several benefits of virtual teams include the following:

- People can work from anywhere at any time.
- People can be recruited for their competencies, not just physical location.
- Many physical handicaps are not a problem.
- Expenses associated with travel, lodging, parking, and leasing or owning a building may be reduced and sometimes eliminated.
- There is no commute time.

7.1.7 REASONS FOR VIRTUAL TEAMS

Reasons for virtual teams center around the differences in time and space for team members.

- Team members may not be physically collocated.
- It may not be practical to travel to meet face to face.
- Team members may work different shifts.

Specifically, teams may be distributed because of the new realities facing organizations such as:

- Organization-wide projects or initiatives
- Alliances with different organizations, some of which may be in other countries
- Mergers and acquisitions
- Emerging markets in different geographic locations
- The desire of many people and government organizations for telecommuting
- The continuing need for business travel and information and communications technologies available to support this travel
- A need to reduce costs
- A need to reduce time-to-market or cycle time in general (the increasing velocity in business)

7.1.8 PRINCIPLES OF VIRTUAL TEAMS AND SYSTEMS THEORY

In their application of systems theory to virtual teams, Lipnack and Stamps (1997) assert that the principles of people, purpose, and links form a simple system model of inputs, processes, and produced outputs. People make up the virtual teams, purpose is the task that holds teams together and links are the interactions and channels that weave the fabric of the team. The nature and variety of these links are the most distinguishing factor between virtual and traditional teams. Figure 1 displays the principles that provide an integrated framework for understanding and working in virtual teams.

The inputs needed to develop virtual teams include independent members, cooperative goals, and multiple media (Lipnack and Stamps, 1997). Throughout the development process, the members share leadership and engage in interdependent tasks, which involve boundary-crossing interactions. The generated outputs include integrated levels of organizations, concrete results, and trusting relationships.

	Inputs	Processes	Produced outputs
People	Independent members	Shared leadership	Integrated levels
Purpose	Cooperative goals	Interdependent tasks	Concrete results
Links	Multiple media	Boundary-crossing interactions	Trusting relationships

(From Virtual Teams, Lipnack & Stamps, 1997).

FIGURE 1:   Virtual Team System of Principles

Virtual teams are composed of individual members with certain areas of expertise. Because of this diversity, members typically share leadership by assuming leadership positions at some point in the process. And because teams are also embedded in organizations, they themselves are parts of larger systems. Therefore, they must integrate both the level of the members and the level of the group.

Three elements of virtual teams allow them to achieve their purpose: cooperative goals, interdependent tasks, and concrete results. Virtual teams rely upon a clear purpose because of their cross-boundary work. Cooperative goals define the outputs desired, while interdependent tasks connect those desired outcomes to those achieved. When a team has completed its process, it expresses its purpose as concrete results.

Links are what give virtual teams their distinction from in-the-same-place organizations. Multiple media (wires, phones, computers, etc.) are the channels by which the members make the physical connection. These connections allow communication and boundary-crossing interaction that make virtual teams truly different. Through interactions, people develop trusting relationships in their patterns of behavior that persist and feed back into subsequent interactions. While it can be argued that trusting relationships are needed by all teams, they are even more important to virtual teams because of a lack of face-to-face time. This trust may even have to replace hierarchical structures and bureaucratic controls (Lipnack & Stamps, 1997).

7.1.9 THE VIRTUOUS LOOP: TEAMS & THE CYBERNETIC MODEL

Cybernetics focuses on the ability of an organization (or team) to engage in self-regulating behavior by a process of negative feedback (Morgan,

1996). By avoiding negative outcomes, or deviations from standard norms, the organization stays on track. The simple cybernetic model, functioning like a thermostat, demonstrates the ability to monitor the environment, as well as the capacity to detect and correct deviations from set guidelines. Modern cybernetics draws the distinction of the ability to question the appropriateness of those predetermined norms before initiating corrective action. Morgan (1997) distinguishes this as the difference between "single-loop" and double-loop learning (Figure 2).

The virtual feedback loop begins with the assumption of a rational model of organization consisting of building blocks of collocated groups stacked in command and control pyramids. Teams work "shoulder to shoulder" and pass their work to the next team in chains of larger processes, similar to a bucket brigade. However, competitive pressures from the environment to cut costs and improve quality are challenging this design. As a consequence, people working on interdependent tasks are no longer necessarily proximate in the space and time or even in the same organization. This leads to problems pertaining to distance, time, and hierarchical structures.

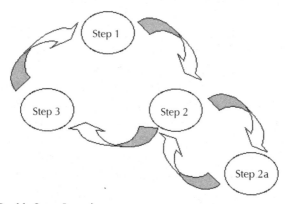

FIGURE 2:   Double-Loop Learning

Step 1   = The process of sensing, scanning, and monitoring the environment

Step 2   = The comparison of this information against other operating norms

Step 2a = The process of questioning whether operating norms are appropriate

Step 3   = The process of initiating appropriate action.

(From Images of Organizations, Morgan, 1996).

Virtual teams address the issue of distance and time by replacing collocation with a combination of technology and face-to-face meetings. They deal with issues pertaining to hierarchical structures through cross-boundary work. This facilitates double-loop learning by creating ways for people to communicate interactively (Geisler, 2002).

Virtual teams are a great way to enable teamwork in situations where people are not sitting in the same physical office at the same time. Such teams are used more and more by companies and other organizations to cut travel, relocation, real estate, and other business costs. This is particularly so for businesses that use virtual organizations to build global presence, outsource their operations, or need less common expertise or skills from people who are reluctant to travel or relocate from their home locations.

Virtual teams are governed essentially by the same fundamental principles as traditional teams. Yet, there is one critical difference. This difference is the way the team members communicate. Instead of using the full spectrum and dynamics of in-office face-to-face exchange, they now rely on special communication channels enabled by modern technologies, such as e-mails, faxes, phone calls and teleconferences, virtual meetings, and alike.

Due to more limited communication channels, the success and effectiveness of virtual teams is much more sensitive to the type of project the group works on, what people are selected, and how the team is managed. Not every type of project is suitable for a virtual organization. One challenging case is projects that rely heavily on sequential or integrated work, as often the case in manufacturing. In particular, when each person's work depends much on what someone else is doing at the same moment (like in a sports team), there is an ongoing heavy exchange of information in real time, and/or the tasks have to go through a strict sequence of workers within a short time.

Not everyone can perform well in a virtual team environment. The members should be self-motivated and able to work independently. They need to be able to keep working effectively without much of external control or structure. The next important quality is strong result orientation. Unless the person shows clear results, there is nobody around to see how intense his or her work activities are. Another critical factor is communication skills. The team member should be able to communicate clearly, constructively, and positively even through the more limited channels of technology, in spite of the loss of many nonverbal cues of face-to-face communications.

Managers of virtual teams also need to pay much more attention to maintaining clear goals, performance standards, and communication rules. People have varying assumptions on what to expect from each other. To avoid build-ups of misunderstandings, in a virtual organization, it is critical to replace those implicit assumptions with clear rules and protocols that everyone understands and agrees upon, especially for communication.

One of the biggest challenges of virtual teams is building and maintaining trust between the team members. Trust is critical for unblocking communication between members and sustaining motivation of each person involved. The issue of trust needs special attention at any stage of team existence (Wellman, 2001).

So, virtual working is whether good opportunity in organization behavior or one of the biggest challenge for future organizations, we need to study those important aspects of organization behavior which are essential for enhancing organization's performance and productivity, that is, perceived organizational climate, leadership effectiveness, self-efficacy, trust, and job involvement.

7.2   PERCEIVED ORGANIZATIONAL CLIMATE

Organizational climate refers to a set of measurable properties of the work environment, that are perceived by the people who live and work in it, and that influence their motivation and behavior. Climate characteristics that have been determined to significantly impact a company's bottom line are: *flexibility, responsibility, standards, rewards, clarity, and team commitment.*

Organizational climate (sometimes known as corporate climate) is the process of quantifying the "culture" of an organization. Organizational cultures are generally deep and stable. Climate, on the other hand, is often defined as the recurring patterns of behavior, attitudes, and feelings that characterize life in the organization (Isaksen & Ekvall, 2007). Although culture and climate are related, climate often proves easier to assess and change. At an individual level of analysis, the concept is called individual psychological climate. These individual perceptions are often aggregated or collected for analysis and understanding at the team or group level, or the divisional, functional, or overall organizational level.

It is a set of properties of the work environment, perceived directly or indirectly by the employees, that is assumed to be a major force in influencing employee behavior.

Reichers and Schneider (1990) defined organizational climate as "the shared perception of the way things are around here" (p. 22). According to Schein (1990), the main differences between organizational climate and organizational culture are the levels of complexity of the two constructs. Organizational culture takes a more in-depth look at the organization's components, whereas organizational climate is simply a surface view of the organization. Denison (1996) reviewed the literature and provided a representation of conceptual differences between the two paradigms. Schein (1990) defined organizational culture as "what a group learns over a period of time as that group solves its problems of survival in an external environment and its problems of internal integration. Such learning is simultaneously a behavioral, cognitive, and an emotional process" (p. 111). Organizational culture was described by Davidson (2003) as "the shared beliefs and values that are passed on to all within the organization" (p. 206). Researchers have shown parallels between the two constructs (Davidson, 2003; Obenchain, 2002; Reichers & Schneider, 1990).

7.2.1   APPROACHES TO DEFINING ORGANIZATION CLIMATE

There are several approaches to the concept of climate, of which two in particular have received substantial patronage: the cognitive schema approach and the shared perception approach.

7.2.1.1   COGNITIVE SCHEMA APPROACH

Cognitive representations of social objects are referred to as schemas. These schemas are a mental structure that represents some aspect of the world. They are organized in memory in an associative network. In these associative networks, similar schemas are clustered together. When a particular schema is activated, related schemas may be activated as well. Schema activation may also increase the accessibility of related schemas in the associative network. When a schema is more accessible, this means it can more quickly be activated and used in a particular situation. When related schemas are activated, inferences beyond the information given in a particular social situation may influence thinking and social behavior, regardless of whether those inferences are accurate or not. Lastly, when a schema is activated, a person may or may not be aware of it.

Two processes that increase the accessibility of schemas are salience and priming. Salience is the degree to which a particular social object stands out relative to other social objects in a situation. The higher the salience of an object, the more likely that schemas for that object will be made accessible. For example, if there is one female in a group of seven males, female gender schemas may be more accessible and influence the group's thinking and behavior toward the female group member. Priming refers to any experiences immediately prior to a situation that caused a schema to be more accessible. For example, watching a scary movie at a theatre late at night might increase the accessibility of frightening schemas that affect a person's perception of shadows and background noises as potential threats.

7.2.2 SHARED PERCEPTION APPROACH

Some researchers have pursued the shared perception model of organizational climate. Their model identifies the variables which moderate an organization's ability to mobilize its workforce in order to achieve business goals and maximize performance.

The major users of this model are departments of the Queensland State Government Australia. These departments use this model of climate to survey staff in order to identify and measure those aspects of a workplace which impact on: stress, morale, quality of work life, well-being, employee engagement, absenteeism/presenteeism, turnover, and performance.

While an organization and its leaders cannot remove every stressor in the daily life of its employees, organizational climate studies have identified a number of behaviors of leaders which have a significant impact on stress and morale. For example, one Queensland state government employer, Queensland Transport, has found that increasing managers' awareness of these behaviors has improved quality of work life employees and the ability of QTs to deliver its organizational goals.

The first approach regards the concept of climate as an individual perception and cognitive representation of the work environment. From this perspective, climate assessments should be conducted at an individual level.

The second approach emphasizes the importance of shared perceptions as underpinning the notion of climate (Anderson & West, 1998; Mathisen & Einarsen, 2004). Reichers and Schneider (1990) define organizational climate as "the shared perception of the way things are around here." It is important to realize that from these two approaches, there is no "best" approach and they actually have a great deal of overlap.

Organizational climate has been defined as the "relatively enduring quality of the internal environment of an organization that a) is experienced by its members, b) influences their behavior, and c) can be described in terms of the values of a particular set of characteristics (or attitudes) of the organization" (Taguiri and Litwin, 1968).

Zammuto and Krackover (1991) measured climate using the following dimensions:

- Autonomy
- Cohesion
- Trust/support
- Pressure
- Recognition
- Fairness
- Innovation

- Credibility
- Conflict
- Trust
- Scapegoating
- Morale
- Equitable rewards
- Resistance to change

Basically, organizational climate reflects a person's perception of the organization to which he belongs. It is a set of characteristics and factors that are perceived by the employees about their organization which serve as a major force in influencing their behavior. These factors may include job description, organizational structural format, performance and evaluation standards, leadership styles, challenges and innovations, organizational values and culture, and so on.

Hodgetts (2006) has classified organizational climate into two major categories. He compares organizational climate with an iceberg where a part of it can be seen above the surface of water and the rest of it is under water and therefore not visible. The visible part of the iceberg that can be observed or measured is akin to the structure of hierarchy, goals, and objectives of the organizations, performance standards and evaluations, technological state of the operations, and so on. This is the first category.

The second category contains factors that are not visible and quantifiable and includes subjective areas such as supportiveness, employee's feeling and attitudes, values, morale, personal and social interaction with peers, subordinates, and superiors, and sense of satisfaction with the job.

Hence, in a nutshell, we can say that

1. Organizational climate is something that is sensed rather than something that is recognized cognitively.
2. Climate is a set of attributes which can be perceived about a particular organization and/or its subsystem deal with their members and environment.
3. Organizational climate is the combined perceptions of individuals that are useful in differentiating organizations according to their procedures and practices.
4. Organizational climate is the collective view of the people within the organization as to the nature of the environment in which they work.

Organizational climate clearly influences the success of an organization. Many organizations, however, struggle to cultivate the climate they need to succeed and retain their most highly effective employees. Hellriegel and Slocum (2006) explain that organizations can take steps to build a more positive and employee-centered climate through:

- **Communication**–how often and the types of means by which information is communicated in the organization
- **Values**–the guiding principles of the organization and whether or not they are modeled by all employees, including leaders
- **Expectations**–types of expectations regarding how managers behave and make decisions
- **Norms**–the normal, routine ways of behaving and treating one another in the organization
- **Policies and rules**–these convey the degree of flexibility and restriction in the organization
- **Programs**–programming and formal initiatives help support and emphasize a workplace climate
- **Leadership**–leaders that consistently support the climate desired

Likert (1967) proposed six dimensions of organizational climate: leadership, motivation, communication, decisions, goals, and control.

Litwin & Stringer (1968) proposed seven dimensions: conformity, responsibility standards, rewards, organizational clarity, warmth and support, and leadership.

Perhaps one of the most important and significant characteristics of a great workplace is its organizational climate. Organizational climate, while defined differently by many researchers and scholars, generally refers to the degree to which an organization focuses on and emphasizes:

- Innovation
- Flexibility
- Appreciation and recognition
- Concern for employee well-being
- Learning and development
- Citizenship and ethics
- Quality performance
- Involvement and empowerment
- Leadership

Organizational climate, manifested in a variety of human resource practices, is an important predictor of organizational success. Numerous studies have found positive relationships between positive organizational climates and various measures of organizational success, most notably for metrics such as sales, staff retention, productivity, customer satisfaction, and profitability (Great workplace, 2009).

7.3   LEADERSHIP

Remember the difference between a boss and a leader: a boss says, "Go!"
- a leader says, "Let's go!"

E. M. Kelly

A leader is best
When people barely know that he exists
Not so good when people obey and acclaim him,
Worst of all when they despise him.
"Fail to honor people,
They fail to honor you;"
But of a good leader, who talks little,

When his work is done, his aim fulfilled,
They will all say, "We did this ourselves".

Lao-Tzu

Leadership is a process whereby an individual influences a group of individuals to achieve a common goal.

Defining leadership as a *process* means that it is not a trait or characteristic that resides in the leader, but a transactional event that occurs between the leader and his or her followers. *Process* implies that a leader affects and is affected by followers. It emphasizes that leadership is not a linear, one-way event, but rather an interactive event. When leadership is defined in this manner, it becomes available to everyone. It is not restricted to only the formally designated leader in a group.

Leadership involves *influence*; it is concerned with how the leader affects followers. Without influence, leadership does not exist.

Leadership occurs in *groups*. Groups are the context in which leadership takes place. Leadership involves influencing a group of individuals who have common purpose. This can be a small task group, a community group, or a large group encompassing an entire organization. Leadership training programs that teach people to lead themselves are not considered a part of leadership within the definition that is set forth.

Leadership includes attention to goals. This means that leadership has to do with directing a group of individuals toward accomplishing some task or end. Leaders direct their energies toward individuals who are trying to achieve something together.

7.3.1 LEADERSHIP DESCRIBED

7.3.1.1 TRAIT VERSUS PROCESS LEADERSHIP

We have all heard statements such as "He is born to be a leader" or "She is a natural leader." These statements are commonly expressed by people who take a trait perspective toward leadership. The trait perspective suggests that certain individuals have special innate or inborn characteristics or qualities that differentiate them from nonleaders. Some of the personal qualities that are used to identify leaders include unique physical factors (e.g., height), personality features (e.g., extraversion), and ability characteristics (e.g., speech fluency) (Bryman, 1992).

To describe leadership as a trait is quite different from describing it as a process. The trait viewpoint conceptualizes leadership as a property or set of properties possessed in varying degrees by different people. This suggests that it resides in select people and restricts leadership to only those who are believed to have special, usually inborn, talents

The process viewpoint suggests that leadership is a phenomenon that resides in the context and makes leadership available to everyone.

A leader is anyone who influences a group toward obtaining a partic-ular result. It is not dependent on title or formal authority (elevos, para-phrased from Leaders, Bennis, and Leadership Presence, Halpern & Lubar). An individual who is appointed to a managerial position has the right to command and enforce obedience by virtue of the authority of his position. However, he must possess adequate personal attributes to match his authority, because authority is only potentially available to him. In the absence of sufficient personal competence, a manager may be confronted by an emergent leader who can challenge his role in the organization and reduce it to that of a figurehead. However, only authority of position has the backing of formal sanctions. It follows that whoever wields personal influence and power can legitimize this only by gaining a formal position in the hierarchy with commensurate authority. Leadership can be defined as one's ability to get others to willingly follow. Every organization needs leaders at every level.

Leaders emerge from within the structure of the informal organization. Their personal qualities, the demands of the situation, or a combination of these and other factors attract followers who accept their leadership within one or several overlay structures. Instead of the authority of position held by an appointed head or chief, the emergent leader wields influence or power. Influence is the ability of a person to gain cooperation from others by means of persuasion or control over rewards. Power is a stronger form of influence because it reflects a person's ability to enforce action through the control of a means of punishment.

Patricia Pitcher (1994) has challenged the bifurcation into leaders and managers. She used a factor analysis (in marketing) factor analysis technique on data collected over 8 years and concluded that three types of leaders exist, each with very different psychological profiles: Artists: imaginative, inspiring, visionary, entrepreneurial, intuitive, daring, and emotional; craftsmen: well-balanced, steady, reasonable, sensible, predict-able, and trustworthy; technocrats: cerebral, detail-oriented, fastidious, uncompromising, and hard-headed. She speculates that no one profile offers a preferred leadership style. She claims that if we want to build, we should find an "artist leader" if we want to solidify our position, we should find a "craftsman leader" and if we have an ugly job that needs to get done like downsizing. We should find a "technocratic leader." Pitcher also observed that a balanced leader exhibiting all three sets of traits occurs extremely rarely: she found none in her study.

Leaders use power and persuasion to motivate followers and arrange the work environment so that they do the job more effectively. Leaders exist throughout the organization, not just in the executive suite.

The competency perspective tries to identify the characteristics of effective leaders. Recent writing suggests that leaders have drive, lead-ership motivation, integrity, self-confidence, above average intelligence, knowledge of the business, and high emotional intelligence.

The behavioral perspective of leadership identified two clusters of leader behavior, people oriented and task oriented.

The contingency perspective of leadership takes the view that effec-tive leadership takes the view that effective leaders diagnose the situation. The path–goal model is prominent contingency theory that identifies four leadership styles–directive, supportive, participative, and achievement–oriented–and several contingencies related to the characteristics of the employee and the situation.

Two other contingency leadership theories are the situational leader-ship model and Fielder's contingency theory. The theory of leadership substitutes identifies contingencies that either limit the leader's ability to influence subordinates or make that particular leadership style unnec-essary. This idea will become more important as organizations remove supervisors and shift toward team-based structures.

Transformational leaders create a strategic vision, communicate that vision through framing and use of metaphors, model the vision by walking the talk and acting consistently, and build commitment to the vision. This leadership style contrasts with transactional leadership, which invokes linking job performance to valued rewards and ensuring that employees have the resources needed to get the job done. The contingency and behav-ioral perspectives adopt the transactional view of leadership.

The crowd will follow a leader who marches twenty steps in advance;
but if he is a thousand steps in front of them, they do not see and do not
follow him.

Georg Brandes

The Conference Board is a not-for-profit organization that conducts research, assesses trends, and makes forecasts about management to help businesses strengthen their performance and better serve society. In 2002, it identified critical skills leaders will need to be successful in the year 2010. Here is the list:

- Cognitive ability: both raw "intellectual horsepower" and mental agility
- Strategic thinking, especially with regard to global competition
- Analytical ability, especially the ability to sort through diverse sources of information and sort out what is most important
- The ability to make sound decisions in an environment of ambiguity and uncertainty
- Personal and organizational communication skills
- The ability to be influential and persuasive with different groups
- The ability to manage an environment of diversity: managing people from different cultures, genders, generations, etc.
- The ability to delegate effectively
- The ability to identify, attract, develop, and retain talented people
- The ability to learn from experience

Source: Barrett, A. and Beeson, J. Developing Business Leaders for 2010. The Conference Board, New York (2002)

7.3.2   LEADERSHIP AND MANAGEMENT

Bennis (1989) stated that leaders inspire and develop others, challenge the status quo, ask what and why questions, and are apt to take a long term view. Managers administer programs, control budgets and costs, maintain the status quo, and are likely to take a short-term view. Building on Bennis's distinctions, one could say that leadership involves changing the way things are, whereas management involves maintaining the current state of affairs.

7.3.3 LEADERSHIP STYLES

Path–goal theory model specifically highlights four leadership styles and several contingency factors leading to three indicators of leadership effectiveness. The four leadership styles are:

7.3.3.1 DIRECTIVE

These are clarifying behaviors that provide a psychological structure for subordinates. The leader clarifies performance goals, the means to reach those goals, and the standards against which performance will be judged. This style also includes judicious use of rewards and disciplinary actions.

7.3.3.2 SUPPORTIVE

These behaviors provide psychological support for subordinates. The leader is friendly and approachable, makes the work more pleasant, treats employees with equal respect, and shows concern for the status, needs, and well-being of the employees. Supportive leadership is the same as people-oriented leadership and reflects the benefits of social support to help employees cope with stressful situations.

7.3.3.3 PARTICIPATIVE

These behaviors encourage and facilitate subordinate involvement in decisions beyond their normal activities. The leader consults with employees, asks for their suggestions, and takes these ideas into serious consideration before making a decision.

7.3.3.4 ACHIEVEMENT-ORIENTED

These behaviors encourage employees to reach their peak performance. The leader sets challenging goals, expects employees to perform their highest level, continually seeks improvement in employee performance, and shows a high degree of confidence that employees will assume responsibility and accomplish challenging goals. Achievement-oriented leadership applies goal-setting theory as well as positive expectations in self-fulfilling prophecy.

The path–goal model contends that effective leaders are capable of selecting the most appropriate behavioral style (styles) for a particular situation. Leaders might use more than one style at a time. For example, they might be both supportive and participative in a specific situation.

7.3.4   LEADERSHIP AT INFOSYS

Successful organizations are built and sustained by great leaders. The transition of leadership at Infosys Technologies when its CEO N. R. Narayana Murthy retired in August 2006 seemed effortless because it was meticulously planned by the company years in advance. Firmly committed to grooming its future leaders, Infosys set up the Infosys Leadership Institute in a sprawling campus in Mysore in 2001.

The aim of the leadership institute is to prepare executives to handle the external and internal business environment and create better customer value through "thought leadership." Chosen leaders, called "high-potential Infoscions" have to undergo a 3-year leadership journey. There are nine pillars of leadership development which have to be imbibed by these Infoscions. The nine pillars are:

1. *360-degree feedback*: To gather data about the individual's performance and abilities
2. *Development assignments*: To train people for various functions and cross-functional assignments
3. *Infosys culture workshops*: To inculcate the Infosys among culture among its employees and develop values and processes involved in leadership development
4. *Developing relationships:* To have one-to-one interaction for better sharing of knowledge with mentoring being part of it
5. *Leadership skills training*: To acclimatize to the next level in the leadership role about the responsibilities of the leader
6. *Feedback intensive programs*: To give formal and informal feedback to employees

7. *Systematic process learning*: To gain an overall view of the company and its diverse and complex systems, business, operations, and processes
8. *Action learning:* To solve problems in real-time conditions in a team
9. *Community empathy*: To stress the need to give back to society through involvement in various developmental, educational, and social causes

7.3.5 THE ROLE OF LEADERS IN SELF-MANAGED WORK TEAMS

When most people think of leaders, they tend to think of individuals who make strategic decisions on behalf of followers and who are responsible for carrying them out. In many of today's organizations, however, where the movement toward *self-managed teams* predominates, it is less likely than ever that leaders are responsible for getting others to implement their orders to help fulfill their visions. Instead, team leaders may be called upon to provide special resources to groups empowered to implement their own missions in their own ways.

Here are a few guidelines that may be followed to achieve success as a team leader.

- Instead of directing people, *team leaders work at building trust and inspiring teamwork.*
- Effective *team leaders concentrate on expanding team capabilities* by functioning primarily as coaches, helping the team by providing all members with the skills needed to perform the task, removing barriers that might interfere with task success, and work at building the confidence of the team, cultivating their untapped potential.
- Instead of managing one on one, *team leaders attempt to create a team identity.*
- *Team leaders are encouraged to make the most of differences between members.* Without doubt, it is a considerable challenge to meld a diverse group of individuals into a highly committed and

productive team, but doing so is important. This can be done by building respect for diverse points of view, making sure that all team members are encouraged to present their views and respecting these ideas once they are expressed.

- Unlike traditional leaders who simply react to change, *team leaders try to foresee and influence change* (Singh, 2010).

7.3.6 CHANGE-ORIENTED LEADERSHIP: FUTURE VISIONS

Companies with the most visionary leaders tend to outperform those with less visionary leaders in all important financial respects.

Having said this, the question arises about what precisely is involved in giving leaders the capacity to envision the most effective changes for the future. Answers are provided by two interesting approaches to leadership known as *charismatic leadership and transformational leadership.*

7.3.6.1 CHARISMATIC LEADERSHIP: "THAT SOMETHING SPECIAL"

Steve Jobs, the pioneer behind Macintosh computer and the growing music download market, has an uncanny ability to create a vision and convince others to become a part of it. This was evidenced by Apple's continual overall success despite its major blunders in the desktop computer wars. Job's ability is so powerful that Apple employees coined a term in the 1980s for it, *the reality-distortion field.* This expression is used to describe the persuasive ability and peculiar charisma of managers like Steve Jobs. This reality-distortion field allows Jobs to convince even skeptics that his plans are worth supporting, no matter how unworkable they may appear. Those close to these managers become passionately committed to seemingly impossible projects, without regard to the practicality of their implementation or competitive forces in the market place. Similarly, people who have worked with Ken Chenault note that they admire him immensely and would do anything for him. He is known for chatting with executives and secretaries alike and is seen as someone who is free from the normal trappings of power (Nelson and Quick, 2009).

7.3.6.1.1 QUALITIES OF CHARISMATIC LEADERS

Researchers have found that charismatic leaders tend to be special in key ways. Specifically, several factors differentiate charismatic leaders from noncharismatic leaders. These are as follows:

- Self-confidence
- A vision
- Extraordinary behavior
- Recognized as change agents
- Environmental sensitivity

7.3.6.2 TRANSFORMATIONAL LEADERSHIP

Kumar Mangalam Birla took over the reins of the Aditya Birla Group after his father, Indian business legend, Aditya Birla, passed away in 1995. The group has scaled several new heights under Birla's transformational leadership. He has won many accolades for his contribution to the industry and efforts to professionalize management (Singh, 2010).

Transformational leadership is about change. It describes a process of positive influence that changes and transforms individuals, organizations, and communities. Transformational leaders influence their constituencies to make the shift from focus on self-interests to a focus on collective interests. They understand the importance of trust building as a means to creating a high commitment to mission-driven outcomes. Research studies have consistently revealed that transformational leadership is positively related to work outcomes. Transformational leadership has been found to be positively related to organizational commitment and job satisfaction (Lussier& Achua, 2011).

7.3.6.2.1 ELEMENTS OF TRANSFORMATIONAL LEADERS

- Transformational leaders are able to set out a bold vision.
- They are skilled at marshalling the intellectual and emotional equity of their people.

- They will not let their heart rule over their mind.
- They encourage intelligence and allow constructive argument.
- They believe in the imperative of institutionalization.
- They are willing to move away from their conventional roles.

To understand transformational leadership, the major components of ethos need to be considered: competence, integrity, likeableness, and dynamism.

The discussion on transformational leadership brings to our attention the transformational CEOs from the industrial and service sectors, such as Jack Welch of GE and Andy Grove of Intel, who are now referred to as management gurus and provide others with extensive advice (Singh, 2009).

Researches surmise that effective transformational leaders:

- See themselves as change agents
- Are visionaries who have a high level of trust for their intuition
- Are risk takers, but not reckless
- Are capable of articulating a set of core values that tend to guide their own behavior
- Possess exceptional cognitive skills and believe in careful deliberation before taking action
- Believe in people and show sensitivity to their needs
- Are flexible and open to learning from experience

7.3.6.3   *STEWARDSHIP AND SERVANT LEADERSHIP*

Stewardship is an employee-focused form of leadership that empowers followers to make decisions and have control over their jobs. Servant leadership is leadership that transcends self-interest to serve the needs of others, by helping them grow professionally and personally. The leader is driven to serve, not to be served. This is similar to qualities of charismatic leaders such as Gandhi (Lussier & Achua, 2011).

7.3.7 CONTEMPORARY LEADERSHIP ROLES

7.3.7.1 MENTORING

A mentor is a senior employee who sponsors and supports a less experienced employee (a protégé). Successful teachers are good teachers. They can present ideas clearly, listen well, and empathize with the problems of their protégés. Mentoring relationships have been described in terms of two broad categories of functions–career functions and psychological functions.

7.3.7.1.1 CAREER FUNCTIONS

- Lobbying to get the protégé challenging and visible assignments
- Coaching the protégé to help develop her skills and achieve work objectives
- Assisting the protégé by providing exposure to influential individuals within the organization
- Protecting the protégé from possible risks to her reputation
- Sponsoring the protégé by nominating her for potential advances or promotions
- Acting as a sounding board for ideas that the protégé might be hesitant to share with her direct supervisor

7.3.7.1.2 PSYCHOLOGICAL FUNCTIONS

- Counselling the protégé about anxieties and uncertainty to help bolster her self-confidence
- Sharing personal experiences with the protégé
- Providing friendship and acceptance
- Acting as a role model

7.3.8 SELF-LEADERSHIP

The underlying assumptions behind self-leadership are that people are responsible, capable, and able to exercise initiative without the external constraints of bosses, rules, or regulations. The importance of self-leadership has increased with the expanded popularity of teams. Empowered, self-managed teams need individuals who are themselves self-directed.

To engage in effective self-leadership: (1) make your mental organizational chart horizontal rather than vertical, (2) focus on influence and not control, and (3) do not wait for the right time to make your mark; create your opportunities rather than wait for them.

7.3.9 THE E-AGE AND ONLINE LEADERSHIP

The present age is often referred to as the electronic age or the e-age. This e-age has added an "e" to virtually everything we see in the present time. But the difference is in the meaning that gets added, which depends on the context. How does this "e" actually make a difference when added to leadership, making it *e-leadership or online leadership?* Business environments in the old industrial age differed from those in today's e-age. Earlier, people used to have 9-to-5 office jobs, relax on weekends, and believe in lifetime employment. Employees usually needed to focus on a single task, and seniority was of utmost importance. In the present age, or the e-age, these concepts have been replaced by that of anytime, anywhere workspace, 24×7 work, play at work, and lifetime learning. Multitasking is important at work, and there is emphasis on knowledge sharing and performance-based working.

The context in which businesses operate today has redefined the role of leaders. How do you lead people who are physically separated from you and with whom your interactions are basically reduced to written digital communications? This is a question that, to date, has received minimal attention from organizational behavior researchers. Leadership research has been directed almost exclusively to face-to-face and verbal situations.

Leaders have to think carefully about what actions they want their digital messages to initiate. Networked communication is a powerful channel. When used properly, it can build and enhance an individual's leadership effectiveness. But when misused, it has the potential to undermine a great deal of what a leader has been able to achieve through his or her verbal actions.

Online leaders confront unique challenges, the greatest of which appears to be developing and maintaining trust. Identification-based trust for instance is particularly difficult to achieve when there is lack of intimacy and face-to-face interaction. In this new world of communications, writing skills are likely to become an extension of interpersonal skills (Robbins, Judge, & Sanghi, 2009).

7.4   SELF-EFFICACY

Perceived self-efficacy is defined as people's beliefs about their capabilities to produce designated levels of performance that exercise influence over events that affect their lives. Self-efficacy beliefs determine how people feel, think, motivate themselves, and behave. Such beliefs produce these diverse effects through four major processes. They include cognitive, motivational, affective, and selection processes.

A strong sense of efficacy enhances human accomplishment and personal well-being in many ways. People with high assurance in their capabilities approach difficult tasks as challenges to be mastered rather than as threats to be avoided. Such an efficacious outlook fosters intrinsic interest and deep engrossment in activities. They set themselves challenging goals and maintain strong commitment to them. They heighten and sustain their efforts in the face of failure. They quickly recover their sense of efficacy after failures or setbacks. They attribute failure to insufficient effort or deficient knowledge and skills which are acquirable. They approach threatening situations with assurance that they can exercise control over them. Such an efficacious outlook produces personal accomplishments, reduces stress, and lowers vulnerability to depression.

In contrast, people who doubt their capabilities shy away from difficult tasks which they view as personal threats. They have low aspirations and weak commitment to the goals they choose to pursue. When faced with difficult tasks, they dwell on their personal deficiencies, on the obstacles they will encounter, and all kinds of adverse outcomes rather than concentrate on how to perform successfully. They slacken their efforts and give up quickly in the face of difficulties. They are slow to recover their sense of efficacy following failure or setbacks. Because they view insufficient performance as deficient aptitude, it does not require much failure for them to lose faith in their capabilities. They fall easy victim to stress and depression.

Psychologist Albert Bandura has defined self-efficacy as our belief in our ability to succeed in specific situations. Your sense of self-efficacy can play a major role in how you approach goals, tasks, and challenges. The concept of self-efficacy lies at the center Bandura's social cognitive theory, which emphasizes the role of observational learning and social experience in the development of personality. According to Bandura's theory, people with high self-efficacy–that is, those who believe they can perform well– are more likely to view difficult tasks as something to be mastered rather than something to be avoided.

In 1977, Bandura developed a theoretical framework for learning and motivation that highlights the role of self-referent thought. Critical to the framework of Bandura's social cognitive theory is the role of self-efficacy (Bandura, 1986). Self-efficacy is defined as one's belief in their ability to perform a given behavior (Bandura, 1977; Wood & Bandura, 1989). Self-efficacy is hypothesized to be an important determinant of action, given the appropriate level of skill and performance (Locke & Latham, 1990) (Spiller, Shane, & Hatfield Robert 2007).

Psychologist Albert Bandura has defined self-efficacy as our belief in our ability to succeed in specific situations. Your sense of self-efficacy can play a major role in how you approach goals, tasks, and challenges. The concept of self-efficacy lies at the center of Bandura's social cognitive theory, which emphasizes the role of observational learning and social experience in the development of personality. According to Bandura's theory, people with high self-efficacy–that is, those who believe they can perform well–are more likely to view difficult tasks as something to be mastered rather than something to be avoided.

Self-efficacy is defined as a person's belief about their ability to organize and execute courses of action necessary to achieve a goal. In other words, persons with strong efficacy beliefs are more confident in their capacity to execute a behavior. Beliefs about self-efficacy have a significant impact on our goals and accomplishments by influencing personal choice, motivation, and our patterns and emotional reactions. For example, we tend to avoid threatening situations that we believe exceed our coping skills. Perceived self-efficacy also affects how successfully goals are accomplished by influencing the level of effort and persistence a person will demonstrate in the face of obstacles. That is, the stronger the perceived self-efficacy, the more active our efforts. Higher self-efficacy is also associated with more persistence, a trait that allows us to gain corrective experiences that reinforce our sense of self-efficacy.

Because of its effect on personal choice, motivation, effort, and persistence, self-efficacy has severe implications for health behaviors, such as condom use and nonsmoking, among others. For example, high self-efficacy influences whether or not a person commits to condom use in the face of social obstacles. Using evidence-based interventions, we can influence self-efficacy through several channels:

1) **Performance Accomplishments** are one's personal mastery experiences, defined as past successes or failures. These experiences form expectations that are generalized to other situations that may be similar or substantially different from the original experience. For example, strong efficacy expectations are developed through repeated success of a behavior, and reduced efficacy expectations can result from failures. We can increase personal mastery for a behavior through participant modeling, performance exposure, self-instructed performances, and performance desensitization, the process through which aversive behavior is paired with a pleasant or relaxing experience.

2) **Vicarious Experience**, which is observing others perform threatening activities without adverse consequences, can also enhance personal self-efficacy by demonstrating that the activity is "doable" with a little effort and persistence. Vicarious experience can be enhanced through live modeling (observing others perform an activity) or symbolic modeling.

3) **Verbal Persuasion**: People are led to believe they can successfully accomplish a task or behavior through the use of suggestion, exhortation, or self-instruction. However, because verbal persuasion is not grounded in personal experience, it is a weaker inducer of efficacy and may be extinguished by histories of past failures.

4) **Emotional Arousal**: We can enhance perceived self-efficacy by diminishing emotional arousals such as fear, stress, and physical agitation since they are associated with decreased performance, reduced success, and other avoidance behaviors. Emotional arousal can be mitigated with repeated symbolic exposure that allows people to practice dealing with stress, relaxation techniques, and symbolic desensitization (the process through which symbolic representation of stressors are paired with a relaxing or pleasant experience).

According to Bandura, a person's attitudes, abilities, and cognitive skills comprise what is known as the self-system. This system plays a major role in how we perceive situations and how we behave in response to different situations. Self-efficacy plays an essential part of this self-system.

Self-efficacy is "the belief in one's capabilities to organize and execute the courses of action required to manage prospective situations" (Bandura, 1995). In other words, self-efficacy is a person's belief in his or her ability to succeed in a particular situation. Bandura (1994) described these beliefs as determinants of how people think, behave, and feel.

Since Bandura published his seminal 1977 paper, "Self-Efficacy: Toward a Unifying Theory of Behavioral Change," the subject has become one of the most studied topics in psychology. Why has self-efficacy become such an important topic among psychologists and educators? As Bandura and other researchers have demonstrated, self-efficacy can have an impact on everything, from psychological states to behavior to motivation.

7.4.1 THE ROLE OF SELF-EFFICACY

Virtually all people can identify goals they want to accomplish, things they would like to change, and things they would like to achieve. However, most people also realize that putting these plans into action is not quite so simple. Bandura and others have found that an individual's self-efficacy plays a major role in how goals, tasks, and challenges are approached.

People with a strong sense of self-efficacy:

- View challenging problems as tasks to be mastered.
- Develop deeper interest in the activities in which they participate.
- Form a stronger sense of commitment to their interests and activities.
- Recover quickly from setbacks and disappointments.

People with a weak sense of self-efficacy:

- Avoid challenging tasks.
- Believe that difficult tasks and situations are beyond their capabilities.
- Focus on personal failings and negative outcomes.
- Quickly lose confidence in personal abilities (Bandura, 1994).

7.4.2   SOURCES OF SELF-EFFICACY

How does self-efficacy develop? These beliefs begin to form in early child-hood as children deal with a wide variety of experiences, tasks, and situations. However, the growth of self-efficacy does not end during youth, but continues to evolve throughout life as people acquire new skills, experiences, and understanding (Bandura, 1992).

According to Bandura, there are four major sources of self-efficacy.

7.4.2.1   MASTERY EXPERIENCES

"The most effective way of developing a strong sense of efficacy is through mastery experiences," Bandura (1994). Performing a task successfully strengthens our sense of self-efficacy. However, failing to adequately deal with a task or challenge can undermine and weaken self-efficacy.

7.4.2.2   SOCIAL MODELING

Witnessing other people successfully completing a task is another important source of self-efficacy. According to Bandura, "Seeing people similar to oneself succeed by sustained effort raises observers' beliefs that they too possess the capabilities master comparable activities to succeed."

7.4.2.3   SOCIAL PERSUASION

Bandura also asserted that people could be persuaded to believe that they have the skills and capabilities to succeed. Consider a time when someone said something positive and encouraging that helped you achieve a goal. Getting verbal encouragement from others helps people overcome self-doubt and instead focus on giving their best effort to the task at hand.

7.4.2.4   PSYCHOLOGICAL RESPONSES

Our own responses and emotional reactions to situations also play an important role in self-efficacy. Moods, emotional states, physical reactions, and

stress levels can all impact how a person feels about their personal abilities in a particular situation. A person who becomes extremely nervous
before speaking in public may develop a weak sense of self-efficacy in
these situations. However, Bandura also notes "it is not the sheer intensity
of emotional and physical reactions that is important but rather how they
are perceived and interpreted" (1994). By learning how to minimize stress
and elevate mood when facing difficult or challenging tasks, people can
improve their sense of self-efficacy.

General self-efficacy is a person's overall view of himself/herself as
being able to perform effectively in a wide variety of situations. Employees
with high general self-efficacy have more confidence in their job-related
abilities and other personal resources (i.e, energy, influence over others,
etc.) that help and may express doubts about performing a new task well.
Previous success or performance is one of the most important determinants of self-efficacy. People who have positive beliefs about their efficacy
for performance are more likely to attempt difficult tasks, persist in overcoming obstacles, and experience less anxiety when faced with adversity.
People with high self-efficacy also value the ability to provide input, or
"voice," at work. Because they are confident in this capability, they value
the opportunity to participate. High self-efficacy has also been recently
related to higher job satisfaction and performance.

There is another form of self-efficacy, called task specific self-efficacy.
Task-specific self-efficacy is a person's belief that he or she can perform
a specific task ("I believe I can do this sales presentation today"). In
contrast, a general self-efficacy is broader ("I believe I can perform well in
just about any part of the job") (Nelson & Quick, 2009).

7.5   TRUST

*Trust is one of the most valuable yet brittle assets in any enterprise. So
over the long term, its far better for companies to downsize in a humane
way.*

Robert Reich

(Mishra, 1998 p.83)

Trust is a relationship of reliance. A trusted party is presumed to seek to
fulfill policies, ethical codes, law, and their previous promises.

Trust is a *prediction* of reliance on an action, based on what a party knows about the other party. Trust is a statement about what is otherwise unknown, for example, because it is far away, cannot be verified, or is in the future.

The degree to which one party trusts another is a measure of belief in the honesty, benevolence, and competence of the other party.

From this perspective, trust is a mental state, which cannot be measured directly. Confidence in the results of trusting may be measured through behavior, or alternatively, one can measure self-reported trust (with all the caveat surrounding that method). Trust may be considered a moral choice, or at least a heuristic, allowing the human to deal with complexities that outgo rationalistic reasoning. In this case, machine–human trust is meaningless, because computers have no moral sense and rely on rational computations. Any trust in a device under this characterization is computer-mediated trust of the user of the machine in the designer and creator of the device, who has implemented the rational rules into the device. Francis Fukuyama (1995) and) are academics who advocate this conception of trust–as moral and not directly observable.

A second perspective in social theory comes from the classic Foundations of Social Theory by James S. Coleman (1990). Coleman offers a four part definition:

1. Placement of trust allows actions that otherwise are not possible (i.e., trust allows actions to be conducted based on incomplete information on the case in hand).
2. If the person in whom trust is placed (trustee) is trustworthy, then the trustor will be better off than if he or she had not trusted. Conversely, if the trustee is not trustworthy, then the trustor will be worse off than if he or she had not trusted (this is reminiscent of a classical prisoner's dilemma).
3. Trust is an action that involves the voluntary placement of resources (physical, financial, intellectual, or temporal) at the disposal of the trustee with no real commitment from the trustee (again prisoner's dilemma).
4. A time lag exists between the extension of trust and the result of the trusting behavior.

In the context of information theory, Gerck (1998) defines and contrasts trust with social functions such as power, surveillance, and accountability.

Trust has been defined as "the willingness of a party to be vulnerable to the actions of another party based on the expectation that the other will perform a particular action important to the trust or, irrespective of the ability to monitor or control that other party" (Mayer, Davis, & Schoorman, 1995).

Trust can be a silent background, sustaining a smooth-running of cooperative relations (Misztal 1996). It can help individuals to reconcile their own interests with those of others. Trust is therefore seen as fundamental to all aspects of social life.

Giddens (1990: 34) defines trust as "confidence in the reality of a person or system, regarding a given set of outcomes or events...." Rousseau, Sitkin, Burt, and Camerer (1998: 394) however claim that there is "no universally accepted scholarly definition of trust."

Giddens however argues that with globalization and the restructuring interactions across undefined spans of time-space, trust which is traditionally secured by community, tradition, and kinship is increasingly vested in abstract capacities characteristic of modern institutions. Trust within organizational groups and teams is a much more complex phenomenon as teams involve multiple, interdependent actors. It is because of this interdependency in team interaction, that some element of trust has to be present for its effective functioning (Jones & George, 1998). Rousseau, Sitkin, Burt, and Camerer (1998) claim that there are different forms of trust. They argue that the "various forms trust can take–and the possibility that trust in a particular situation can mix several forms together–account for some of the apparent confusion among scholars."

Modern scholars trying to bring together issues of trusted systems, computer security, trust and technology include Jeroen van den Hoven, Helen Nissenbaum, Deborah Johnson, Jean Camp, and Ed Gerck.

Trust is an essential element in leadership. Trust is the willingness to be vulnerable to the actions of another. This means that followers believe that their leader will act with the follower's welfare in mind. Trustworthiness is also one of the competencies in emotional intelligence. Trust among top management team members facilitates strategy implementation; this means that if team members trust each other, they have a better chance of getting "buy in" from employees on the direction of the company. And if employees trust their leaders, they will buy in more readily.

How would you go about leading a team of people in different organizations, in different geographic locations around the world, who have never met? They would not have shared understandings of problems, norms, work distribution, roles, or responsibilities. This is a challenge that is becoming more common, and one that Boeing-Rocketdyne faced. What Boeing-Rocketdyne learned is that the leader of such teams needs to be the "spoke in the center of the wheel" in terms of coordination. The leader also needs to help the team create a common language and document results for the entire team. Not surprisingly, Boeing's largest rival–Airbus Industries of Europe–has developed its own virtual teams. Called Elab, this network helps Airbus coordinate work by aerospace firms all over Europe, including British Aerospace, Rolls Royce, and Snecma. Using complex communication tools, including high quality video, Elab allow these member firms to create complete working environments for groups of engineers scattered throughout the continent. Leading virtual teams require trust, because face-to-face interaction that is the hallmark of leadership is not possible. Leaders must not only come to trust their subordinates, but they must also express that trust. Research has shown that workers who believe their boss trusts them (called "felt trustworthiness") enjoy their work more, are more productive, and are more likely to "go the extra mile" at work and perform organizational citizenship behaviors.

Effective leaders also understand both, *who* to trust and *how* to trust. At one extreme, leaders often trust a close circle of advisors, listening only to them and gradually cutting themselves off from dissenting opinions. At the opposite extreme, lone wolf leaders may trust nobody, leading to preventable mistakes. Wise leaders carefully evaluate both the competence and the position of those they trust, seeking out a variety of opinions and input (Nelson & Quick, 2009).

Conventional wisdom assumes that trust develops from a history of interpersonal interactions and communication, through which people come to "know and trust" one another. In virtual teams, however, establishing trust can be complicated: members may have no past on which to build, no future to reference, and may never even actually meet face to face. Swift but fragile trust can develop early in a team's life cycle. Yet, if swift trust does not develop or even dissipates, members need to find ways of building trust in each other. To this end, an understanding of how trust impacts a virtual team's development will help managers and team leaders to facilitate and improve team success. Herein, we describe the three components of trust (ability, integrity, and benevolence) and identify which of these are critical to each life cycle stage (establishing the team, inception, organizing, transition, and accomplishing the task) of the virtual team. Proposed action steps for each stage show managers and team (Greenberg & Antonucci, 2007).

According to Robbins (2011), trust is a positive expectation that another will not act opportunistically. He identified five dimensions that underlie the concept of trust–integrity, competence, consistency, loyalty, and openness. He considers trust as the foundation for leadership. Three types of trust have been identified in organizational relationships:

- Deterrence-based trust: Trust based on fear of reprisal if the trust is violated and most fragile.
- Knowledge-based trust: Trust based on behavioral predictability that comes from a history of interaction.
- Identification-based trust: Trust based on a mutual understanding of each other's intentions and appreciation of the other's wants and desires.

7.5.1   SYMPTOMS OF LACK OF TRUST IN ORGANIZATIONS

Some of the warning signs that trust is an issue in the organization are:

- An active, inaccurate grapevine
- Elaborate approval processes
- Low initiative
- High turnover
- High fear factor among employees
- Turf wars
- Defensiveness

A leadership guru, management consultant, and successful author Tom Peters (1993) captures the predicament this way:

Maybe the boss can force a person to show up for work, especially in trying times, but one cannot by definition force a person to contribute his or her passion and imagination on a regular basis.

According to Jack R. Gibb (1991), the reasons for lack of trust and defensiveness among individuals are the fear of top management, intolerable pressures on employees, crisis situations, lack of clarity of vision among employees, employee pressure on management, and basic unrest in organizations. As a result of this fear and distrust, a culture of dependence, passivity, and excessive conformity is likely to develop in organizations resulting in a climate of low initiative.

7.5.2 STEPS FOR BUILDING TRUST IN ORGANIZATIONS

There are four steps involved in building trust:

- Effective communication: If the communication is open and genuine and does not involve hidden agendas, it does not leave any space for confusion and doubt.
- Real understanding: If the first step has been followed, it leads to real understanding between people as a result of the honest and deep sharing of thoughts and ideas.
- Development of relationship of mutual respect: A relationship of respect demands reciprocity of respect from each other. If the respect is unilateral, it leads to a very superficial and temporary relationship. A state of mutual respect comes from effective understanding and binds people together and facilitates long-term commitment.
- Maintaining the integrity of the relationship: Integrity involves adhering to a code of ethics or a set of values. It is a vital aspect of every personal and professional endeavor.

7.6 JOB INVOLVEMENT

Job involvement refers to your involvement with or alienation from a specific job.

Lodahl and Kejner (1965) have defined the term job involvement as the internalization of value about goodness of work or importance of work in the worth of the person; perhaps it thus measures the case with which the person can be further socialized by an organization. Job-involved employees are likely to believe in the work ethic to exhibit high-growth decision-making.

Job involvement has been defined as an individual's psychological identification or commitment to his/her job (Kanungo, 1982). It is the degree to "which one is cognitively preoccupied with, engaged in, and concerned with one's present job" (Paullay et al., 1994, p. 224). Job involvement involves the internalization of values about the goodness of work or the importance of work in the worth of the individual (Lodahl & Kejner, 1965). As such, individuals who display high involvement in their jobs consider their work to be a very important part of their lives and whether or not they feel good about themselves is closely related to how they perform on their jobs. In other words, for highly involved individuals, performing well on the job is important for their self-esteem (Lodahl & Kejner, 1965). Because of this, people who are high in job involvement genuinely care for and are concerned about their work (Kanungo, 1982).

The construct of job involvement is somewhat similar to organizational commitment in that they are both concerned with an employee's identification with the work experience. However, the constructs differ in that job involvement is more closely associated with identification with one's immediate work activities, whereas organizational commitment refers to one's attachment to the organization (Brown, 1996). It is possible, for example, to be very involved in a specific job but not be committed to the organization or vice versa (Blau & Boal, 1987).

Research studies over the past two decades, which have explored the construct of job involvement, have approached it from two different perspectives (Sekeran, 1989; Sekeran & Mowday, 1981). First, when viewed as an individual difference variable, job involvement is believed to occur when the possession of certain needs, values, or personal characteristics predispose individuals to become more or less involved in their jobs. For instance, Rabinowitz and Hall (1977) in their review of literature on job involvement found that individual characteristics such as age, education, sex, tenure, need strength, level of control, and values were linked to job involvement. The second perspective views job involvement as a response to specific work situation characteristics. In other words, certain types of jobs or characteristics of the work situation influence the degree to which an individual becomes involved in his/her job.

For example, research has demonstrated that job involvement has been related to job characteristics such as task autonomy, task significance, task identity, skill variety, and feedback and supervisory behaviors such as leader consideration, participative decision-making, and amount of communication (Brown, 1996).

Job involvement (employee engagement, or work engagement, is a concept that is generally viewed as managing discretionary effort, that is, when employees have choices, they will act in a way that furthers their organization's interests. An engaged employee is a person who is fully involved in, and enthusiastic about, his/her work. In his book, *Getting Engaged: The New Workplace Loyalty*, author Tim Rutledge explains that truly engaged employees are attracted to, and inspired by, their work ("I want to do this"), are committed ("I am dedicated to the success of what I am doing"), and are fascinated ("I love what I am doing").

Kahn was the first scholar to define "personal engagement" as the "... harnessing of organization member's selves to their work roles: in engagement, people employ and express themselves physically, cognitively, emotionally, and mentally during role performances" (p.694). Based on this definition, a questionnaire was developed that assesses three dimensions: cognitive, emotional, and physical engagement.

An alternative academic considers work engagement as a psychological state of fulfillment and the positive antithesis of burnout. It is defined as "...a positive, fulfilling, work-related state of mind that is characterized by vigor, dedication, and absorption." Whereby vigor is characterized by high levels of energy and mental resilience while working, the willingness to invest effort in one's work, and persistence even in the face of difficulties; dedication by being strongly involved in one's work, and experiencing a sense of significance, enthusiasm, inspiration, pride, and challenge; and absorption by being fully concentrated and happily engrossed in one's work, whereby time passes quickly and one has difficulties with detaching oneself from work.

Job involvement is how people see their jobs as both a relationship with the working environment, the job itself, and how their work and life are commingled. Having low job involvement contributes to employees' feelings of alienation of purpose, alienation in the organization, or feeling of separation between what the employees see as their life and the job they do. Work alienation and job involvement are correlated with one another (Hirschfeld & Field, 2000)

It represents the extent to which an individual is personally involved with his or her work role. Kanungo (Work Alienation, 1982) defines job involvement as the psychological identification with one's job.

There are many definitions of job involvement. Job involvement is defined as the degree to which a person psychologically identifies with. It is related with the work motivation that a person has with a job (Bashaw & Grant, 1994; Hackett et al., 2001; McElroy et al., 1995; Blau, 1986; Blau & Boal, 1987; Balay, 2000).

Job involvement is the internalization of values about the work or the importance of work according to the individual. Job involvement may appraise the ease with which a person can be further socialized by an organization. Organizational socialization is the process by which an individual understands the values, abilities, behaviors, and social knowledge indispensable for an organizational role and for taking part in as a member (Ramsey et al., 1995).

It is a belief about one's current job and is a function of how much the job can satisfy one's wishes. Highly job-involved individuals make the job a central part of their personal character. Besides, people with high job involvement focus most of their attention on their jobs (Hackett et al., 2001).

Job involvement is grouped into four diverse categories: (1) work as a central life interest, (2) active participation in the job, (3) performance as central to self-esteem, and (4) performance compatible with self-concept. In work as a central life interest, job involvement is thought of as the degree to which a person regards the work situation as important and as central to his/her identity because of the opportunity to satisfy main needs. In active participation in the job, high job involvement hints the opportunity to make job decisions, to make an important contribution to company goals, and self-determination. Active participation in the job is thought to ease the achievement of such needs as prestige, self-respect, autonomy, and self-regard. In performance as central to self-esteem, job involvement implies that performance on the job is central to his/her sense of worth (Ramsey et al., 1995; Blau & Boal, 1987; Balay, 2000).

Job involvement is a function of individual difference and the work situation. Thus, demographic and work experience variables are expected to relate to job involvement. Positive relationships are expected with age, tenure, years in occupation, education, having children, and gender. There is no evidence for a strong relationship between job involvement and performance (Cohen, 1999).

Job involvement is negatively associated with intentions to quit and positively related to job satisfaction and organizational climate perceptions (McElroy et al., 1995; McElroy et al., 1999). In the same way, Blau and Ryan put forward that (1997: 437) job involvement and organizational commitment are negatively related to absence, withdrawal intentions, and turnover as well as lateness and leaving work early, and job involvement is positively related to work effort and performance. Individuals with high levels of both job involvement and organizational commitment should be the most motivated to go to work and to go on time. Individuals with low levels of job involvement and organizational commitment should be the least motivated. Both highly motivated and nonmotivated employees may miss work or come late for excusable reasons (e.g., illness, religious holiday, vacation time, and transportation problems). However, highly motivated employees cannot be thought as nonmotivated employees to miss work or come late for inexcusable reasons. Individuals with higher levels of job involvement and organizational commitment are likely to exhibit less unexcused lateness and unexcused absence than individuals with lower levels of job involvement and organizational commitment (Blau, 1986;

Blau and Boal (1987) portray four different groups to classify employees by: (a) high job involvement–high organizational commitment; (b) high job involvement–low organizational commitment; (c) low job involvement–high organizational commitment; and (d) low job involvement–low organizational commitment. Employees in the first group are labeled "institutionalized stars," in the second group "lone wolves," in the third group "corporate citizens," and in the fourth group "apathetic employees."

There has also been some research into organizational commitment and job involvement especially related to the health-care workers and nurses (Brewer & Lok, 1995; Brooks & Swailes, 2002; Örs et al., 2003; Özsoy et al., 2004; Sjöberg & Sverke, 2000; Blau & Boal, 1989).

In a study conducted by Sjöberg and Sverke (2000) in Swedish Emergency Hospital, it was found out that organizational commitment and job involvement had a variety of consequences on turnover. Blau and Boal (1989) found that nurses with higher levels of job involvement and organizational commitment had significantly less unexcused absences than nurses with lower levels of job involvement and organizational commitment. Yet, no study into organizational commitment and job involvement of the state employees at Ministry of Health has been performed.

KEYWORDS

- globalization
- knowledge management
- virtual team
- cybernetics
- leadership
- self-efficacy

CHAPTER 8

REVIEW OF LITERATURE

CONTENTS

The literature of investigation is carried out in the field of organizational climate, leadership, self-efficacy, interpersonal trust, and job involvement.

8.1   ORGANIZATIONAL CLIMATE

Putti & Kheun,1986 examined the relationship between organizational climate and job satisfaction in one of the departments in the Civil Service in Singapore. While organizational climate was conceptualized as a characteristic of organizations which is reflected in the descriptions employees make of the policies, practices, and conditions which exist in the environment; job satisfaction refers to affective orientations on the part of individuals toward work dimensions. The sample includes professional, technical, and administrative people. The data collected through climate and job satisfaction instruments was subjected to correlation analysis by using Pearson Product Moment Formula. The overall finding of the study was that job satisfaction was highly correlated with organizational climate.

A two-phase study was conducted on the relationship between organizational climate variables and burnout among personnel in a multifunction community service agency. Initial interviews with a small pilot sample identified major sources of stress, which were then incorporated into a second-phase questionnaire to all staff. Multiple regression analyses illustrated that perceived interactions between head office administrators and sections of the agency contributed significantly to emotional exhaustion, whereas perceptions of within-section interactions and involvement in decision-making had a positive impact on personal accomplishment. Perceived communication levels, however, were negatively related to personal accomplishment (O'driscoll & Schubert, 1988).

A total of 865 bank employees from all grades completed the Spector Work Locus of Control Scale and the adapted Furnham Corporate Assessment Scale, which measure corporate climate on various dimensions. Both questionnaires demonstrated high internal reliability and were closely related. The results suggest that internal locus of control is related to a positive perception of organizational climate, particularly commitment and morale (Furnham & Drakeley, 1993).

Schneider (1996) found that service and performance climates predict customer satisfaction.

In an exploratory study : Virtual Teams vs Face to Face Teams by Merrill, Warkentin, Sayeed & Hightower (1996), it was found that teams using computer mediated communication system (CMCS) could not outperform traditional (face-to-face) teams under otherwise comparable circumstances. Further relational links among team members were found to be a significant contributor to the effectiveness of information exchange. Though virtual and face-to-face teams exhibited similar levels of communication effectiveness, face-to-face team members report higher levels of satisfaction.

Lipnack and Stamps (1997) stated that managing a successful virtual company requires 90% people and 10% technology. A virtual manager is faced with far more challenges of keeping members connected and communicating effectively across the network. Policies and procedures are also established and necessary for members to follow and respect. Management reluctance can also be a problem for virtual corporations. Not everyone embraces the virtual team work model around the world. Some developing country's managers may like a traditional office instead of a virtual corporation

Young and Parker (2000) examined the extent to which collective climates are comprised of individuals with similar interpretive schemata such as work values and need strength, or consist of individuals who share work group or interaction group membership. Measures of psychological climate, work values, need strength, and employee interaction patterns were collected from the management and administrative staff of a manufacturing organization. Results supported the symbolic interactionist perspective to the formation of collective climates. It was found with clear evidence that collective climates are related to employee interaction groups. Employee interaction based on sense making and information seeking activities was most strongly related to shared climate perceptions. There was also some evidence that individuals with similar levels of need, strength share collective climate membership.

Potosky and Ramakrishna (2001) found that an emphasis on learning and skill development was significantly related to organizational performance.

Organizational identification, which reflects how individuals define themselves with respect to their organization, may be called into question in the context of virtual work. Virtual work increases employees' isolation and independence, threatening to fragment the organization. It was found that virtual workers' need for affiliation and the work-based social support they experience are countervailing forces associated with stronger organizational identification. Furthermore, perceived work-based social support moderates the relationship between virtual workers' need for affiliation and their strength of organizational identification. Thus, when work-based social support is high, even workers with lower need for affiliation may strongly identify with the organization.

Organizational identification is important in a virtual setting because it may replace or otherwise compensate for the loss of aspects of traditional organizations that facilitate cooperation, coordination, and the long-term effort of employees. For example, when employees can work anytime and anywhere, it is difficult to rely upon mechanisms, such as direct supervision as a means of coordination and control (DeSanctis & Monge, 1999). Instead, it may be left to the discretion of employees themselves to self-organize, being motivated to seek out, and provide cooperative behaviors (e.g., organizational citizenship behaviors) that further task performance and organizational goals. Organizational identification, or the strength of members' psychological link to the organization, has been associated with the degree to which employees are motivated to fulfill organizational needs and goals, their willingness to display organizational citizenship and other cooperative behaviors, and their tendency to remain with the organization (Dutton, Dukerich, & Harquail, 1994; Kramer, 1993; Mael & Ashforth, 1995). Overall, our ability to manage the large and growing population of virtual employees may depend on identifying the factors that predict their organizational identification (Wisenfeld, Raghuram, & Garud, 2001).

Before Pettigrew's (1979) landmark organizational culture study, research centered on the construct of organizational climate (Obenchain, 2002).

There has been increasing interest in the field of customer retention in the last two decades. Much of that interest has focused on the economics of customer retention and developing plans and strategies for companies to follow in order to improve customer retention. There has been little research into what determines customer retention, particularly from the perspective of organizational climate. The relationship between employees' perception of organizational climate and customer retention in a specific service setting, namely a major UK retail bank was examined. Employees' perceptions of the practices and procedures in relation to customer care at their branch were investigated using a case study approach. The findings revealed that there is a relationship between employees' perceptions of organizational climate and customer retention at a microorganizational level. They suggested that organizational climate can be subdivided into five climate themes and that, within each climate theme, there are several dimensions that are critical to customer retention (Clark, 2002).

Cooper and Kurland (2002) employed a grounded theory methodology to compare the impact telecommuting has on public and private employees perceptions of professional isolation. It relied on 93 semistructured interviews with telecommuters, nontelecommuters, and their respective supervisors in two high technology firms and two city governments. These organizations had active telecommuting programs and a strong interest in making telecommuting a successful work option, providing an opportunity to investigate the challenges of telecommuting that existed even within friendly environments. The interviews demonstrated that professional isolation of telecommuters was inextricably linked to employee development activities (interpersonal networking, informal learning, and mentoring). The extent to which telecommuters experience professional isolation depended upon the extent to which these activities were valued in the workplace and the degree to which telecommuters missed these opportunities. Public respondents appeared to value these informal developmental activities less than private employees. Therefore, it was stipulated that telecommuting is less likely to hinder the professional development of public sector employees than that of employees in the private sector.

Using training history data and supervisory ratings of 133 hazardous waste workers' safety performance collected within two organizations in the US nuclear waste industry, Crowe, Burke, and Landis (2003) examined organizational climate for the transfer of safety training as a moderator of relationships between safety knowledge and safety performance. The trend in the results was consistent with the hypothesis that these relationships would be stronger in the less restrictive (more supportive) organizational climate.

Employees who were attending classes at a local university responded to measures of perceived organizational support, the content of their psychological contracts (e.g., relational and transactional obligations), social and economic exchange, the level of fulfillment of both employee and organizational obligations, and organizational commitment. Part-time employees ($N = 319$) reported higher levels of perceived organizational support and stronger economic exchange relationships, while full-time employees ($N = 282$) reported higher levels of continuance commitment, sacrifice, and greater relational and transactional obligations to their organizations. There were no significant differences between the two groups in terms of the strength of social exchange relationships, the levels of their organizations' relational and transactional obligations to them, the degree to which they had fulfilled their obligations to their organizations or their organizations had fulfilled their obligations to them, the level of continuance commitment, perceived alternatives, affective commitment, and normative commitment. There were no differences in the strength of the relations between perceived organizational support and the other exchange variables depending on work status. Overall, the findings suggested that social exchange processes operate similarly for part-time and full-time employees (Gakovic & Tetric, 2003).

Patterson, Warr, and West (2004) found that manufacturing organizations that emphasized a positive organizational climate, specifically concern for employee well-being, flexibility, learning, and performance, showed more productivity than those that emphasized these to a lesser degree.

A number of studies by Rose and colleagues (2001, 2002, 2004) have found a very strong link between organizational climate and employee reactions, such as stress levels, absenteeism and commitment, and participation. A study by Heidi Bushell (2007) has found that Hart, Griffin et al.'s (1996) organizational climate model accounts for at least 16% single-day sick leave and 10% separation rates in one organization. Other studies support the links between organizational climate and many other factors, such as employee retention, job satisfaction, well-being, and readiness for creativity, innovation and change. Hunter, Bedell, and Mumford (2007) have reviewed numerous approaches to climate assessment for creativity. They found that those climate studies that were based on well-developed, standardized instruments produced far higher effect sizes than did studies that were based on locally developed measures.

Organizational climate emphasizes the importance of shared perceptions as underpinning the notion of climate (Anderson & West, 1998; Mathisen & Einarsen, 2004).

Seven dimensions of organizational climate and measures of perceived customer satisfaction were gathered from food and beverage employees of 14 hotels. Regression analysis revealed organizational climate to explain 26.9% of the variance in customer satisfaction with food and beverage and only two organizational climate dimensions, professional and organizational esprit, and conflict and ambiguity, displaying a unique relationship to customer satisfaction with food and beverage. Customer satisfaction with food and beverage was found to explain 18.45% of the variation in among the hotels. Recommendations were made as to which dimensions of organizational climate should be targeted for intervention programs attempting to increase hotel financial performance (Davidson & Manning, 2004).

Baba, Gluesing, Ratner, and Wagner (2004) focused on cognitive convergence in a GDT, which is defined as the process by which cognitive structures of distributed team members gradually become more similar over time. To explore the convergence process, a longitudinal, ethnographic research strategy was employed that allowed to follow a naturally occurring globally distributed team, over a 14-month period, producing a rich case study portraying factors and processes that influenced convergence. Confirming previous studies, it was found that increases in shared cognition alone are not sufficient to account for performance gains on a globally distributed team. Rather, it may be necessary not only to increase the sharing of cognition, but also to reverse a pattern of increasing divergence that can result from rejection of key knowledge domains. It was found that several factors influence the process of cognitive convergence beyond direct knowledge sharing. These included separate but parallel or similar learning experiences in a common context, the surfacing of hidden knowledge at remote sites by third-party mediators or knowledge brokers, and shifts in agent self-interest that motivate collaboration and trigger the negotiation of task interdependence. Also relevant to cognitive convergence on a GDT is the geographical distribution pattern of people and resources on the ground, and the different ways in which leaders exploit the historical, cultural, and linguistic dimensions of such distribution to further their own political agendas. It was concluded that GDTs can be effective in bringing together divergent points of view to yield new organizational capabilities, but such benefits require that leaders and members recognize early and explicitly the existence and validity of their differences.

Organizational Climate has positive effect on employee's knowledge sharing or organizational learning (Bock, Zmud, Kim, & Lee, 2005; Janz & Prasarnphanich, 2003; Zhang & Huang, 2005)

Henttlonen and Blomqeut (2005) argued that information technology plays an important role in virtual teams but virtual team work involves significant social redesign.

Adopting a person's environment fit perspective, Fletcher, Major, and Davis (2007) examined the influence of competition as an interaction between trait competitiveness and competitive climate. Using a sample of information technology workers, competitive climate was considered as both an individual-level variable and a work group variable. Results show that the effect of competitive climate depended on trait competitiveness and the level at which climate was assessed for four of the outcomes assessed: job satisfaction, organizational commitment, job dedication, and supervisor-rated task performance. In general, the effect of competitive climate was more negative for individuals lower in trait competitiveness. Competitive psychological climate was associated with greater stress regardless of the level of trait competitiveness but was not directly related to self-rated task performance. Findings suggest that managers should be cautious in encouraging competitive climate.

Findings of 957 surveyed employees from four evangelical higher education institutions found a negative correlation for climate and commitment and staff members. Administrators were found to have a more favorable view of their institutional climate than staff. Employee age, tenure, and classification had predictive value for organizational climate, whereas only employee age and tenure predicted organizational commitment (Thomas, 2008).

Recently, climate strength, which is an index of the level of agreement within a group around perceptions of climate, has been shown to have a moderating effect on the relationship between climate level and organizational outcomes. Sowinski, Fortmann, and Lezotte (2008) studied and made a contribution to the emerging body of climate strength research as it attempted to constructively replicate the findings of Schneider, Salvaggio, and Subirats (2002) with a sample of 756 employees from 129 stores within the automotive services industry, and expanded the organizational outcome variables to include employee turnover and profitability. Support was not found for the hypothesis that climate strength would moderate the relationship between climate levels and organizational outcomes. However, significant main effects were found for some of the climate strength–outcome relationships.

Jordan, Leon, Epstein, Durkin, Helgerson, and Lakin (2009) examined the association between organizational climate and changes in internalizing and externalizing behavior for youth in residential treatment centers (RTCs). The sample included 407 youth and 349 front-line residential treatment staff from 17 residential treatment center in Illinois. Youth behavior was measured using the Child Functional Assessment Rating Scale. Organizational climate was measured via the Areas of Work-life Survey. Using hierarchical linear modeling, results demonstrated that a higher perception of person-job match on community among front-line staff was associated with more improvement on youth externalizing behaviors. Counterintuitively, higher person-job match on fairness and workload were each associated with less improvement on internalizing and externalizing behavior.

Kunze, Boehm, and Bruch (2010) investigated (a) the effect of organizational-level age diversity on collective perceptions of age discrimination climate that (b) in turn should influence the collective affective commitment of employees, which is (c) an important trigger for overall company performance. In a large-scale study that included 128 companies, a total of 8651 employees provided data on their perception of age discrimination and affective commitment on the company level. Information on firm-level performance was collected from key informants. They tested the proposed model using structural equation modeling (SEM) procedures and, overall, found support for all hypothesized relationships. The findings demonstrated that age diversity seems to be related to the emergence of an age discrimination climate in companies, which negatively impacts overall firm performance through the mediation of affective commitment.

Perceptions of organizational climate, leadership, and group processes were aggregated within hierarchically nested work groups. Relationships across hierarchical boundaries were examined for two samples at different hierarchical levels in a military organization. Perceptions of climate were positively related across levels in both samples. There was evidence that the pattern of relationship among the other constructs was different in the two samples (Griffin & Matheiu, 2010).

Time pressure, one of the factors of organizational innovation climate, has an inconsistent effect on employee creativity. Based on the interactional approach, this study attempted to describe time pressure as a moderator. Data were collected from two surveys of R&D employees at Taiwanese national research institutions in 2007 and 2009. The results showed that time pressure moderated the relationship between organizational innovation climate and creative outcomes. As most theorists had predicted, in a strong organizational innovation climate, time pressure hindered creative outcomes. However, as many practitioners advocate, time pressure enhanced creative outcomes in a weak organizational innovation climate (Hsu & Fan, 2010).

A number of studies by Dr Dennis Rose and colleagues between 2001 and 2004 have found a very strong link between organizational climate and employee reactions, such as stress levels, absenteeism and commitment, and participation.

A study has found that Hart, Griffin et al.'s (1996) organisational climate model accounts for at least 16% single-day sick leave and 10% separation rates in one organization. Other studies support the links between organizational climate and many other factors, such as employee retention, job satisfaction, well-being, and readiness for creativity, innovation, and change. Hunter, Bedell, and Mumford have reviewed numerous approaches to climate assessment for creativity. They found that those climate studies that were based on well-developed, standardized instruments produced far higher effect sizes than did studies that were based on locally developed measures.

8.2   LEADERSHIP

Yukl (1994) pointed out that leadership effectiveness, like leadership, was defined differently by different researchers. Most of them consider the results produced by employees under the influence of the leader's behavior as an indicator in the assessment of leadership effectiveness, including employees' satisfaction with the superior, employees' identity with organizational objectives, employees' development and psychological health, and superior's promotion opportunity.

Sosik and Godshlk (2000) examined linkages between mentor leadership behaviors (laissez-faire, transactional contingent reward, and transformational), protégé perception of mentoring functions received (career development and psychosocial support) and job-related stress of 204 mentor–protégé dyads. Results of partial least squares analysis revealed that mentor transformational behavior was more positively related to mentoring functions received than transactional contingent reward behavior, while mentor laissez-faire behavior was negatively related to mentoring functions received. Both mentor transformational behavior and mentoring functions received were negatively related to protégé job-related stress. The relationship between mentor transformational behavior and protégé job-related stress was moderated by the level of mentoring functions received.

Jung and Avolio (2000) examined the causal effects of transformational and transactional leadership and the mediating role of trust and value congruence on followers' performance. A total of 194 student participants worked on a brainstorming task under transformational and transactional leadership conditions. Leadership styles were manipulated using two confederates, and followers' performance was evaluated via three measures–quantity, quality, and satisfaction. Results, based on path analyses using LISREL, indicated that transformational leadership had both direct and indirect effects on performance mediated through followers' trust in the leader and value congruence. However, transactional leadership had only indirect effects on followers' performance mediated through followers' trust and value congruence.

According to Reilly & Ryan, 2000 initial evidence suggest that as teams become more virtual the ambassadorial behaviors will have positive influence on the necessary conditions for a high-performing team. These include trust, organizational citizenship behaviors, innovative thinking, and shared mental models. The ambassadorial leadership model provides a framework for the development of practical leadership skills. As with most leadership behaviors, these skills can be learned and mastered with practice. Those organizations that deploy virtual teams in either operational or strategic initiatives may find that training in these skills will reap rewards in team effectiveness and also in improved levels of organizational citizenship behavior among their employees.

On the basis of the current theories of charismatic leadership, several possible follower effects were identified. It was hypothesized that followers of charismatic leaders could be distinguished by their greater reverence, trust, and satisfaction with their leader and by a heightened sense of collective identity, perceived group task performance, and feelings of empowerment. Using the Conger–Kanungo charismatic leadership scale and measures of the hypothesized follower effects, an empirical study was conducted on a sample of 252 managers using structural equation modeling. The results showed a strong relationship between follower reverence and charismatic leadership. Follower trust and satisfaction, however, were mediated through leader reverence. Followers' sense of collective identity and perceived group task performance were affected by charismatic leadership. Feelings of empowerment were mediated through the followers' sense of collective identity and perceived group task performance (Conger, Kanungo, & Menon, 2000).

Crant and Bateman (2000) tested hypotheses regarding the relationship between proactive personality and perceptions of charismatic leadership. A sample of 156 managers completed measures of proactive personality along with measures of the five-factor model of personality and other individual differences. The managers' immediate supervisors rated their charismatic leadership and in-role behavior. Results suggested that self-reported proactive personality is positively associated with supervisors' independent ratings of charismatic leadership. Hierarchical regression analyses revealed that proactive personality accounts for variance in a manager's charismatic leadership above and beyond that accounted for by an array of control variables (the Big Five personality factors, in-role behavior, and social desirability).

Multilevel marketing organizations (MLMs) are a rapidly growing organizational type enlisting nearly 10 million members and producing over 20 billion dollars in sales annually. Despite their remarkable recent growth, few studies have examined these unusual organizations, and none of these have addressed issues of transformational leadership. In Multilevel marketing organizations, the key leadership relationships are those between individual member distributors and the members who recruited them into the organization (i.e., their "sponsors"). Although sponsors are expected to provide leadership to the members they recruit, they possess no direct supervisory resulting–authority in an uncertain "quasi-leadership" role. Using a sample of 736 female multilevel marketing organization members, the present study empirically tests an important explanatory component of transformational leadership theory: that belief in the higher purpose of one's work is a mechanism through which transformational leadership achieves its positive outcomes on cohesion, satisfaction, effort, and performance. The results offered support to the notion that transformational leadership indeed "transforms" followers by encouraging them to see the higher purposes in their work. Additionally, the results showed positive relationships between belief in a higher purpose of one's work and job satisfaction, unit cohesion, and effort (Sparks & Schenk, 2001).

Koene, et.al (2002) examined the effect of different leadership styles on two financial measures of organizational performance and three measures of organizational climate in 50 supermarket stores of a large supermarket chain in the Netherlands. Findings showed a clear relationship of local leadership with financial performance and organizational climate in the stores. The findings also showed that the leadership styles have differential effects. Charismatic leadership had a substantial effect on climate and financial performance in small stores.

Using questionnaire and interview data, Sagie, Zaidman, Amichai-Hamburger, Te'eni, and Schwartz (2002) attempted to find out whether the organizational loose (participative) and tight (directive) practices are compatible with or contradict each other. Using the theoretical framework of Sagie's loose–tight leadership approach, the hypotheses concerned the effects of both practices on the employee's work-related attitudes, and the mediating role of two variables, cognitive (information sharing) and motivational (exerting effort), in these effects. Data were analyzed using two methodological approaches, quantitative and qualitative. Based on a quantitative analysis of the questionnaires given to 101 professional employees of a textile company, partial support was provided for the study hypotheses. A qualitative analysis of in-depth, semi-structured interviews with all the employees ($n = 20$) in one of the company divisions led to similar conclusions. Specifically, it was found that although the loose and tight practices affected work attitudes, the interviewees attributed more impact to the tight practice. In addition, none of the study variables mediated the loose impact on attitudes, whereas information sharing (but not exerting effort) mediated the influence of tight practice. Finally, the qualitative analysis revealed a deeper insight into the nature of both leader practices and their possible integration in the decision-making processes in organizations. In the face of heightened competitive pressures, elevated quality expectations, and calls for worker empowerment, more and more organizations have turned to self-directed work teams (SDWTs). A review of the literature devoted to self-directed work teams suggests that managers often struggle with the transition to self-directed work teams because of the required shift in control to self-directed work teams members. To promote the development of work teams, managers must modify their use of influence tactics in direct response to the control shift.

Mohammed, Mathieu, and Bart (2002) examined the mix of ability, experience, and personality impacts three types of team performance: Technical-administrative task performance, leadership task performance, and contextual performance. Relationships were tested using data collected from student management teams, who were required to plan and supervise the preparation and service of meals in a cafeteria-style dining room patronized by university students, staff, and faculty. Results revealed that both team- and task-related composition variables predicted leadership and contextual performance. Specifically, grade-point average was significantly related to technical-administrative task performance, and extraversion, neuroticism, and grade-point average were related to leadership task performance. Agreeableness and restaurant experience predicted contextual performance. Surprisingly, conscientiousness did not account for significant variance in any of the three types of performance measured.

Youngjin and Maryam (2003) conducted an exploratory study to examine the behaviors and role that are enacted by emergent leaders in virtual teams settings. The longitudinal study involved seven ad hoc and temporary virtual teams composed of senior executives of a US federal government agency. It was found that overall, the emergent leaders sent more and longer email messages than their team members did. The number of task-oriented messages, particularly those that were related to logistics, coordination, sent by emergent leaders, was higher than that of nonleaders. Furthermore, the emergent leaders enacted three roles: Initiator, scheduler, and integrator.

Relationships of participative leadership with relational demography variables (age, tenure, education, and gender) were explored in an integrated model, combining the average leadership style (ALS) and the leader–member exchange (LMX) approaches to leadership. Data were collected from 561 staff members from 36 schools. The RWG and the WABA (within- and between-analysis) results indicated the prevalence of the LMX model and the individual-differences approach in explaining the relationship of the leader's participative behaviors with relational demography variables. In addition, consistent with the study hypotheses, the negative relationship between demographic dissimilarity and participative decision-making (PDM) was stronger in short-term superior–subordinate relationships than in longer-term relationships (Somech, 2003).

Cable and Judge (2003) question: Why do managers employ certain tactics when they try to influence others? by proposing and studying theoretical linkages between the five-factor model of personality and managers' upward influence tactic strategies. Longitudinal data from 189 managers at 140 different organizations confirmed that managers scoring high on extraversion were more likely to use inspirational appeal and ingratiation; those scoring high on openness were less likely to use coalitions; those scoring high on emotional stability were more likely to use rational persuasion and less likely to use inspirational appeal; those scoring high on agreeableness were less likely to use legitimization or pressure; and those scoring high on conscientiousness were more likely to use rational appeal. Results also confirmed that managers' upward influence tactic strategies depended on the leadership style of their target (their supervisor). Managers were more likely to use consultation and inspirational appeal tactics when their supervisor was a transformational leader, but were more likely to use exchange, coalition, legitimization, and pressure tactics when their supervisor displayed a laissez-faire leadership style.

Garman, Davis & Corrigan, 2003 examined the factor structure of the transformational leadership model in human service teams. As the nature of this work environment mandates certain management-by-exception practices, patterns of correlations between perceptions of active and passive management-by-exception behaviors and transformational, transactional, and laissez-faire leadership were of interest. A total of 236 leaders and 620 subordinates from 54 mental health teams completed the Multifactor Leadership Questionnaire, form 8Y. Results suggest that active and passive management-by-exception factors are independent constructs.

Using a sample of 520 staff nurses employed by a large public hospital in Singapore, Avolio, Zhu, Koh, and Bhati (2004) examined whether psychological empowerment mediated the effects of transformational leadership on followers' organizational commitment. They also examined how structural distance (direct and indirect leadership) between leaders and followers moderated the relationship between transformational leadership and organizational commitment. Results from HLM analyses showed that psychological empowerment mediated the relationship between transformational leadership and organizational commitment. Similarly, structural distance between the leader and follower moderated the relationship between transformational leadership and organizational commitment.

Hill (2004) examined and found that due to geographic dispersion and reliance on technology-mediated communication, developing collaborative capital can be a challenge in a virtual team. Knowledge sharing is one form of collaborative capital that has been identified as critical to virtual team success. She proposed that shared leadership in virtual teams is positively related to knowledge sharing between team members and this relationship is partially mediated by trust.

Changes in managers' usage of influence tactics during the transition to SDWTs within a large aluminum manufacturing plant were explored.

Analyses of longitudinal data show that despite the new team environment, managers' use of influence tactics was focused at the individual level. It was also found that transition time accounts for variance in managers' choices of influence tactics. Finally, an exploratory analysis suggested that high as opposed to low self-monitoring managers may be more prone to increase their usage of soft influence tactics and decrease their usage of hard influence tactics over the course of the transition; the influence behavior of low self-monitoring managers remained unchanged (Douglas & Gardner, 2004).

Spreitzer, Pertula, and Xin (2005) examined how the effectiveness of transformational leadership may vary depending on the cultural values of an individual and developed the logic for why the individual value of traditionality (emphasizing respect for hierarchy in relationships) moderates the relationship between six dimensions of transformational leadership and leadership effectiveness. The hypotheses were examined on leaders from Asia and North America. The results indicate support for the moderating effect of traditional values on the relationship between four dimensions of transformational leadership (appropriate role model, intellectual stimulation, high performance expectations, and articulating a vision) on leadership effectiveness.

Laura, Thomas, O'Neill, and Theresa (2005) studied the effects of transformational and transactional leadership styles and communication media on team interactional styles and outcomes in which teams communicated through one of the following three ways: (a) face to face, (b) desktop video conference, or (c) text-based chat. They found that transformational and transactional leadership styles did not affect interaction styles or outcomes; that the mean constructive interaction score was higher in face-to-face than videoconference and chat team.

Using longitudinal data collected in two waves, 9 months apart, from 372 employees, Boomer, Rich & Rubin,2005 studied empirical assessment of individual-level change within an organizational setting. Specifically, strategies used by change implementers were operationalized as six transformational leader behaviors, and then hypothesized to influence employees' cynicism about organizational change (CAOC). A combination of social learning theory, and communication research served as the theoretical rationale to explain transformational leadership's hypothesized effects. As posited, transformational leader behaviors (TLB) generally were associated with lower employee CAOC. Further, the direction of causality was consistent in suggesting that the TLB reduced employee CAOC.

Noncognitive emotional intelligence could potentially contribute to a more holistic understanding of interpersonal influence and leadership; however, significant issues of definition, psychometric independence, and measurement must be conclusively resolved (Brown & Moshavi, 2005).

Hambley, A O'Neill, and Kline (2005) studied the effects of transformational and transactional leadership styles and communication media on team interaction styles and outcomes. Teams communicated through one of the following three ways: (a) face to face, (b) desktop videoconference, or (c) text-based chat. Results indicated that transformational and transactional leadership styles did not affect team interaction styles or outcomes; that the mean constructive interaction score was higher in face-to-face than videoconference and chat teams, but not significantly higher in videoconference than chat teams; and that teams working in richer communication media did not achieve higher task performance than those communicating through less rich media. Finally, mean team cohesion scores were higher in face-to-face and videoconference than chat teams, but not significantly higher in face-to-face than videoconference teams. These results provide further evidence that communication media do have important effects on team interaction styles and cohesion.

Challenges associated with leading a $1.7 trillion industry have created a need for strong leaders at all levels in health-care organizations. However, despite growing support for the importance of leadership development practices across industries, little is known about leadership development in health-care organizations. An extensive qualitative study comprising 35 expert interviews and 55 organizational case studies included 160 in-depth, semistructured interviews and explored this issue. Across interviews, several themes emerged around leadership development challenges that were particularly salient to health-care organizations. Informants described how the relative newness of leadership development practices in a majority of health-care organizations contributes to an overall perception of haphazard practices throughout the industry. In addition, respondents noted challenges associated with developing leaders who would be representative of the patient community served, and commented on the pressure to segregate different professional groups for leadership development. Framed by these challenges, a conceptual model of commitment to leadership development in health-care organizations as influenced by three factors–strategy, culture, and structure–was proposed. This can, in turn, influence program design decisions and can impact organizational effectiveness (McAlearney, 2006).

Hmieleski and Ensley (2007) examined the relationship of entrepreneur leadership behavior (empowering and directive), top management team heterogeneity (functional, educational specialty, educational level, and skill), and industry environmental dynamism (rate of unpredicted change in number of industry establishments, number of industry employees, industry revenue, and industry research and development intensity) on new venture performance (revenue growth and employment growth) using two different samples–the Inc. 500 list of America's fastest growing startups and a national (United States) random sample of new ventures. In dynamic industry environments, startups with heterogeneous top management teams were found to perform best when led by directive leaders and those with homogenous top management teams performed best when led by empowering leaders. Conversely in stable industry environments, startups with heterogeneous top management teams were found to perform best when led by empowering leaders and those with homogenous top management teams performed best when led by directive leaders. These findings were consistent across both samples and demonstrate the value in a contextual approach to leadership, which considers adjusting leadership behavior in accordance to factors that are both internal and external to the firm.

Euwema, Wendt, and Emmerik (2007) investigated (a) the effects of societal culture on group organizational citizenship behavior (GOCB) and (b) the moderating role of culture on the relationship between directive and supportive leadership and group organizational citizenship behavior. Data were collected from 20,336 managers and 95,893 corresponding team members in 33 countries. Multilevel analysis was used to test the hypotheses, and culture was operationalized using two dimensions of Hofstede (2001) and GLOBE (2004): individualism and power distance (PD). There was no direct relationship between these cultural dimensions and group organizational citizenship behavior. Directive leadership had a negative relation, and supportive leadership a positive relation with group organizational citizenship behavior. Culture moderated this relationship: directive leadership was more negatively, and supportive behavior less positively, related to group organizational citizenship behavior in individualistic compared to collectivistic societies.

Schneier and Bartol (2007) studied sex effects in emergent leadership by assigning 52 task groups for 15 weeks course in personnel administration randomly to 108 female and 176 male undergraduates. Results showed no significant differences in the proportion of males and females to emerge as leaders through sociometric choice.

Spreitzer (2007) looked at how the leadership practices of business organizations may foster more peaceful societies and developed the logic for positive relationships between participative organizational leadership, employee empowerment, and peace. Several mechanisms to explain why these different manifestations of voice are likely to contribute to peaceful societies were offered. Support for the hypotheses regarding the positive effects of participative leadership and employee empowerment in work organizations on peace was found.

Ogbeide and Cho (2008) studied to determine if managers' perceptions of their leadership styles are in agreement with their subordinates' perceptions, and to determine which of the four leadership styles are influential in predicting employees' performance. The result of this study indicates disagreement between managers' and subordinates' perceptions of the managers' leadership styles. Self-other agreement on leadership styles predicted employees' performance. Supportive, participative, and achievement-oriented leadership styles were influential in predicting employees' motivation. Supportive and directive leadership styles increased employees' intention to get the job done.

Berson, Oreg, and Dvir (2008) examined the relationships between CEO values and organizational culture, and between organizational culture and firm performance. Data were collected from different sources (26 CEOs, 71 senior vice presidents, and 185 other organizational members), and include organizational financial performance data collected at two points in time. In support of hypotheses, CEO self-directive values were associated with innovation-oriented cultures, security values were associated with bureaucratic cultures and benevolence values were related to supportive cultures. In turn, cultural dimensions showed differential associations with subsequent company sales growth, an index of organizational efficiency and assessments of employee satisfaction.

Walumbwa, Luthans, Avey, and Oke (2009) examined at the group level of analysis the role that collective psychological capital and trust may play in the relationship between authentic leadership and work groups' desired outcomes. Utilizing 146 intact groups from a large financial institution, the results indicated a significant relationship between both their collective psychological capital and trust with their group-level performance and citizenship behavior. These two variables were also found to mediate the relationship between authentic leadership and the desired group outcomes, even when controlling for transformational leadership.

Utilizing self-complexity theory and other aspects of research on self-representation, Hannah, Woolfolk & Lord, 2009 showed how the structure and structural dynamics of leaders' self-constructs are linked to their varied role demands by calling forth cognitions, affects, goals and values, expectancies, and self-regulatory plans that enhance performance. Through this process a leader was able to bring the "right stuff" (the appropriate ensemble of attributes) to bear on and succeed in the multiple challenges of leadership.

Hirst and Knippenberg (2009) used a social identity analysis to predict employee creativity and hypothesized that team identification leads to greater employee creative performance, mediated by the individual's creative effort. They hypothesized that leader inspirational motivation as well as leader team prototypicality would moderate the relationship between identification and creative effort. Consistent with these predictions, data based on 115 matched pairs of employee–leader ratings in a research and development context showed an indirect relationship between team identification and creative performance mediated by creative effort. The analyses also confirmed the expected moderated relationships. Leader inspirational motivation enhanced the positive association between identification and creative effort, especially when leader prototypicality was high.

Innovative behavior is increasingly important for organizations' survival. Transformational leadership, in contrast to transactional leadership, has been argued to be particularly effective in engendering follower innovative behavior. However, empirical evidence for this relationship is scarce and inconsistent. Addressing this issue, it was proposed that follower's psychological empowerment moderates the relationship of transformational and transactional leadership with follower innovative behavior. In a field study, with 230 employees of a government agency in the Netherlands combining multisource ratings, it was shown that transformational leadership is positively related to innovative behavior only when psychological empowerment is high, whereas transactional leadership has a negative relationship with innovative behavior only under these conditions (Pieterse, van Knippenberg, Schippers, & Stam, 2010).

It is generally argued that leader visions motivate followers by focusing on reaching desirable end states. However, it has also been suggested that visions may motivate followers by focusing on avoiding undesirable situations. Stam, Knippenberg, and Wisse (2010) investigated the effects of appeals that focus on preventing an undesirable situation (i.e., prevention appeals) as well as appeals that focus on promoting a desirable situation (i.e., promotion appeals) and argue that the effectiveness of promotion and prevention appeals is contingent on follower regulatory focus. In two experiments it was shown that prevention-appeals lead to better performance than promotion-appeals for more prevention-focused followers, while the reverse is true for more promotion-focused followers. This pattern was found for a dispositional measure of follower regulatory focus (Study 1), as well as for a manipulation of follower regulatory focus (Study 2).

Using the LMX theory as a theoretical framework, van Breukelen, van der Leeden, Wesselius, and Hoes (2010) focused on the occurrence of differential treatment by leaders on social and task-related issues within teams. It was investigated whether team members' perceptions of the frequency and degree of social and task-related differential treatment by the leader were associated with their evaluation of team atmosphere and team performance, in addition to the effects of the quality of their own working relationship with the leader (LMX quality). The context of this study consisted of interdependent sports teams. The participants were 605 players belonging to 69 amateur sports teams playing various team sports such as soccer, hockey, and basketball. Social differential treatment was negatively associated with team atmosphere and unrelated to team performance. In addition, it was found that the two forms of task-related differential treatment included in this study were unrelated to team atmosphere and were differently associated with team performance.

8.3 SELF-EFFICACY

Self-efficacy (one's belief in one's capability to perform a task) affects task effort, persistence, expressed interest, and the level of goal difficulty selected for performance. Despite this, little attention has been given to its organizational implications (Gist, 1987).

Little research has been done on characteristics of successful self-managed work group members, despite the fact that almost every major

US corporation is considering implementing such teams. Thoms, Moore, and Scott (1996) examined the relationship between the Big Five personality dimensions and self-efficacy for participating in self-managed work groups. A questionnaire was administered to 126 workers in a manufacturing organization that is planning the implementation of self-managed work groups. Results indicated that neuroticism, extraversion, agreeableness, and conscientiousness were significantly related to self-efficacy for participating in self-managed work groups. Due to the relationship between self-efficacy and performance, one implication of these findings is that organizations should consider personality when deciding whether or not to implement self-managed work groups or who should be selected to work in this type of structure.

A key foundation of empowering organizations is employee self-leadership. (Prussia, Anderson, & Manz, 1998) examined the effects of self-leadership skills and self-efficacy perceptions on performance. Structural equations modeling determined whether the influence of self-leadership on performance is mediated by self-efficacy perceptions. Results for the sample of 151 respondents indicated that self-leadership strategies had a significant effect on self-efficacy evaluations, and self-efficacy directly affected performance. Further, self-efficacy perceptions were found to fully mediate the self-leadership/performance relationship.

Staples, Hulland, and Higgins (1999) investigated how virtual organizations can manage remote employees effectively. The research used self-efficacy theory to build a model that predicts relationships between antecedents to employees' remote work self-efficacy assessments and their behavioral and attitudinal consequences. The model was tested using responses from 376 remote managed employees in 18 diverse organizations. Overall, the results indicated that remote employees' self-efficacy assessments play a critical role in influencing their remote work effectiveness, perceived productivity, job satisfaction, and ability to cope. Furthermore, strong relationships were observed between employees' remote work self-efficacy judgments and several antecedents, including remote work experience and training, best practices modeling by management, computer anxiety, and IT capabilities.

Given that self-efficacy has been shown to be positively related to training outcomes, a better understanding of factors that affect self-efficacy in complex training contexts was needed. The development of self-efficacy in a flight-training program was examined. Results indicate that training performance and self-esteem predicted self-efficacy for post-training flight performance. Furthermore, prior flight experience moderated the relationships between training performance and self-efficacy, and between self-esteem and self-efficacy (Davis, Fedor, Parsons, & Herold, 2000).

Efficacy–effectiveness relationships were examined for individual nurses and nursing teams who were either trained or untrained in goal setting. At the individual level positive direct relationships were demonstrated between self-efficacy and effectiveness, between training and subsequent self-efficacy, and between training and effectiveness. Nurses low in initial self-efficacy realized greater effectiveness gains from the training than did nurses high in initial self-efficacy. At the team level, group efficacy was related to effectiveness, and training was related to subsequent group efficacy but it was not related to effectiveness and there were no moderation effects for initial group efficacy (Gibson, 2001).

Raghuram, Wiesenfeld, Batia, and Garud (2001) explored factors associated with employees' ability to cope with the challenges of telecommuting—an increasingly pervasive new work mode enabled by advances in information technologies. Telecommuting can trigger important changes in employees' job responsibilities, especially with respect to the degree of proactivity required to effectively work from a distance. Survey responses from a sample of 723 participants in one organization's formal telecommuting program were used to examine the interrelationships between telecommuter self-efficacy and extent of telecommuting on telecommuters' ability to cope with this new work context. Results indicate that there is a positive association between telecommuter self-efficacy and both employees' behavioral strategies (i.e., structuring behaviors) and work outcomes (i.e., telecommuter adjustment).

Paglis and Green (2002) developed and tested a leadership model that focused on managers' motivation for attempting the leadership of change. The construct of leadership self-efficacy (LSE) was defined, and a measure comprising three dimensions (direction setting, gaining followers' commitment, and overcoming obstacles to change) was developed. Based on Bandura's (1986) social cognitive theory, the primary hypothesis is that high leadership self-efficacy managers will be seen by direct reports as engaging in more leadership attempts. Relationships were also proposed between leadership self-efficacy and several factors that are expected to influence this confidence judgment. Managers' organizational commitment and crisis perceptions were modeled as potential moderators of the relationship between leadership self-efficacy and leadership attempts. The model was tested through surveys distributed to managers ($n = 150$) and their direct reports ($n = 415$) in a real estate management company and an industrial chemicals firm. Positive relationships ($P < 0.05$) were found between the first two dimensions of leadership self-efficacy and managers' leadership attempts. An interaction effect involving organizational commitment was discovered for the leadership self-efficacy/overcoming obstacles dimension ($P < 0.05$). Several positive relationships were found between LSE dimensions and proposed antecedents, including self-esteem ($P < 0.05$), subordinates' performance abilities ($P < 0.05$), and managers' job autonomy ($P < 0.05$).

Self-efficacy belief is a significant predictor of behavioral choices in terms of goal setting, the amount of effort devoted to a particular task, and actual performance. The study conceived of formation and change of self-efficacy as a social and context-dependent process, and hypothesized that different group factors (discretionary and ambient group stimuli) influence changes in members' self-efficacy through differing routes (individual-level and cross-level processes). Hypotheses was tested using data from individuals in 169 training groups who attended a 5-day workshop designed to increase participants' job-search skills and efficacy. Specifically, the degree of change in participants' job-search efficacy before and after the workshop was examined. The results showed that (a) membership diversity in education was positively related to increases in job-search efficacy, (b) supportive leadership contributed to job-search efficacy at the individual level of analysis with no cross-level effects, and (c) open group climate contributed to job-search efficacy through both individual-level and cross-level processes (Choi, Price, & Vinokur, 2003).

New business formation is a formidable and daunting task, which may require personal perseverance and self-efficacy. If this is indeed the case, will entrepreneurs and non-entrepreneurs differ on such attributes? Also, if high levels of perseverance and self-efficacy help entrepreneurs to overcome setbacks, snags, and obstacles, do these positive attributes cooccur with significant personal costs, such as the tendency to experience regretful thinking? Markman, Baron, and Balkin (2005) used a random sample of 217 patent inventors in the medical industry (surgery devices) to address these questions. Results indicated that entrepreneurs score significantly higher on self-efficacy and on two distinct aspects of perseverance–

perceived control over adversity and perceived responsibility regarding outcome of adversity–than did nonentrepreneurs. Also, although entrepreneurs reported the same number of regrets, their regrets were stronger and were qualitatively different from those reported by nonentrepreneurs. These findings suggested that perseverance and self-efficacy do indeed cooccur with regretful thinking. Finally, posthoc analysis revealed that the higher the overall perseverance scores of patent inventors, the higher their annual earnings.

Staples, Higgins, and Hulland (2006) indicated that remote employee's self-efficacy assessment play a critical role in influencing their remote work effectiveness, perceived productivity, job satisfaction, and ability to cope. Furthermore, strong relationships were observed between employee's remote work as self-efficacy judgments and several antecedents, including remote work experience and training, best practices modeling by management, computer anxiety, and IT capabilities.

Nauta, Lin, and Chaoping (2006) evaluated the cross-national validity of cognitive appraisal theories (e.g., Lazarus and Folkman, 1984) of stress by examining differences in the interactions of job autonomy and generalized self-efficacy in the prediction of psychological and physical strains among US and Chinese employees. It was found that for Chinese employees with high self-efficacy, job autonomy was negatively related to job strains, but for Chinese employees with low self-efficacy, job autonomy was positively related to job strains.

Luthans, Norman, Avolio, and Avey (2008) investigated whether the recently emerging core construct of positive psychological capital (consisting of hope, resilience, optimism, and efficacy) plays a role in mediating the effects of a supportive organizational climate with employee outcomes. Utilizing three diverse samples, results showed that employees' psychological capital was positively related to their performance, satisfaction, and commitment and a supportive climate is related to employees' satisfaction and commitment.

8.4 TRUST

Current research suggests that virtual teams failure is directly related to the difficulties of building trust, positive relationships across the three boundaries of geographical distances, time zones, and cultural differences (Kimble et al., 2000).

Raguram and Wiensfeld (2004) studied work–nonwork conflict and job stress among virtual workers and found preliminary evidence suggesting that virtual workers are negatively related to work–nonwork conflict and job stress. It was also found that work factors (clarity of appraisal criteria, interpersonal trust, and organizational connectedness) and individual factors (self-efficacy and ability to structure the workday) associated with work–nonwork conflict and found that these associations are moderated by the extent of virtual work.

8.4.1 SWIFT TRUST IN TEMPORARY TEAMS

After the team has begun to interact, trust is maintained by a "highly active, proactive, enthusiastic, generative style of action" (Meyerson et al., 1996). High levels of action have also been shown to be associated with high-performing teams (Lacono & Weisband, 1997). Action strengthens trust in a self-fulfilling fashion: action will maintain member's confidence that the team is able to manage the uncertainty, risk, and points of vulnerability, yet the conveyance of action has as a requisite the communication of individual activities. In summary, whereas traditional conceptualizations of trust are based strongly on interpersonal relationships, swift trust deemphasizes the interpersonal dimensions and is based initially on broad categorical social structures and later on action. Since members initially import trust rather than develop trust, trust might attain its zenith at the projects inception (Meyerson et al., 1996).

Developed to explain behavior in temporary teams, such as film crews, theater and architectural groups, presidential commissions, senate select committees, and cockpit crews (Meyerson et al., 1996); the theory of swift trust assumes clear role divisions among members who have well-defined specialties. Inconsistent role behavior and "blurring" of roles erode trust. Moreover, the theory seems to presuppose that participants come from many different organizations, have periodic face-to-face meetings, and report to a single individual. By contrast, in global virtual teams, members remain in different locations and often are accountable to different individuals. Such teams are assembled less based on their specific roles and more based on their knowledge differences, partially related to the geographic location of the individual, which provides them with greater knowledge of that environment. These differences may have significant implications for swift trust. In the temporary teams described by Meyerson et al. (1996), what is at stake are the professional reputations of members, the reputations of the persons to whom the team members report, impending threats from closely knit social and professional groups to which members and the supervisor belong, and perceived interdependence among the team members. In global virtual teams, the reputational and professional network effects may be weak because of less clearly defined and bounded professional networks and less emphasis on roles Clark and Payne (1997) presented a theoretical and empirical analysis of the nature of trust at work. Use was made of the facet approach to generate a definitional framework of trust, and it was proposed as a theoretical basis for the analysis of the structural characteristics of trust. Hypotheses regarding the relations between the definitional framework and empirical observations were tested by applying smallest space analysis to analyze data collected from a sample of 398 colliery workmen, using questionnaires developed on the basis of the faceted definition. The results demonstrated strong support for the definitional framework suggested for the concept of trust and its construct validity. The results also suggested a possible distinction workers make between trust and mistrust and, between the specifics of activities to do with the job itself vs managers in general.

Early work on trust in the virtual environment has found that short-lived teams are in fact able to develop high trust but they do so by following a swift trust model rather than the traditional model of trust development (Jarvenpaa et al., 1998; Jarvenpaa & Leidner, 1999).

The swift trust paradigm suggests that when they do not have enough time to slowly build trust, team members assume that others are trustworthy and begin working as if trust were already in place while seeking confirming or disconfirming evidence throughout the duration of the project (Meyerson et al., 1996). Virtual teams that exhibit high trusting behaviors experience significant social communication as well as predictable communication patterns, substantial feedback, positive leadership, enthusiasm, and the ability to cope with technical uncertainty (Jarvenpaa & Leidner, 1999).

It was examined organizational citizenship of residents in a housing cooperative setting where roles were not influenced by traditional employee–employer work relationships. Results demonstrate that the individual differences of collectivism and propensity to trust predicted organizational citizenship (assessed 6 months later). In addition, organizational-based self-esteem fully mediated the effects of collectivism and propensity to trust on organizational citizenship, and tenure moderated the trust–self-esteem relationship (Van Dyne, Vandewalle, Kostova, Latham, & Cummings, 2000).

Trust development in virtual teams also presents significant challenges because it is difficult to assess teammates' trustworthiness without ever having met them (McDonough et al., 2001). Moreover, as the life of many virtual teams is relatively limited, trust must quickly develop (Jarvenpaa & Leidner, 1999). Yet, trust development is deemed crucial for the successful completion of virtual team projects (Sarker et al., 2001).

This early research identified perceived integrity of other team members as particularly important in the development of trust early in a team's life and perceptions of other members' benevolence as a trait that supported the maintenance of trust over time (Jarvenpaa et al., 1998). High-trust teams may also develop as a result of early face-to-face meetings with the intent of developing a strong foundation of trust between members (Suchan & Hayzak, 2001), or thanks to communication training (Warkentin & Beranek, 1999).

Data obtained from full-time employees of a public sector organization in India were used to test a social exchange model of employee work attitudes and behaviors. LISREL results revealed that whereas the three organizational justice dimensions (distributive, procedural, and interactional) were related to trust in organization, only interactional justice was related to trust in supervisor. The results further revealed that relative to the hypothesized fully mediated model a partially mediated model better fitted the data. Trust in organization partially mediated the relationship between distributive and procedural justice and the work attitudes of job satisfaction, turnover intentions, and organizational commitment but fully mediated the relationship between interactional justice and these work attitudes. In contrast, trust in supervisor fully mediated the relationship between interactional justice and the work behaviors of task performance and the individually- and organizationally-oriented dimensions of citizenship behavior (Aryee, Budhwar, & Chen, 2002).

Trust is of great interest in organizational research and the social sciences. It is argued here that any theory of trust ought to treat a set of observed empirical regularities as constraints. A constraint on a theory is either an assumption made in constructing the theory (e.g., we assume people have a tendency to be fair in their dealings with others) or an explanandum (e.g., we must explain the tendency people have to be fair in their dealings with others) for which the theory must account. These constraints are best thought of in an evolutionary framework, with intellectual links to game theory, evolutionary biology and evolutionary psychology. In particular, (1) people are disposed toward fairness and reciprocity, they have (2) cognitive modules, and (3) emotional dispositions which safeguard them against cheating and trust violation. Given these three constraints, any theory should consider the (4) evidence for cultural constraints on trust, trusting behavior, and decisions. These constraints together point toward an integrated ecology of trust where culture, individual dispositions, and cognition jointly influence observed trusting behavior. Some work, particularly Gintis et al. (2002) on altruism, strong reciprocity and cooperation, suggests such an approach to studying trust. The purpose of this essay is to identify this work, organize it, and argue for its bearing on the study of trust and organizations (Fichman, 2003).

Virtual teams represent a new form of organization that offers unprecedented levels of flexibility and responsiveness, and has the potential to revolutionize the workplace. Virtual teams, however, cannot be implemented on faith and they do not represent an organizational panacea (Powell, Piccoli, & Blake, 2004).

Webber, Webbe, and Klimoski (2004) study addressed the behaviors of project managers under various conditions of cognitive and affective trust, and the implications for obtaining client loyalty. Theoretical foundations were drawn from interpersonal and interorganizational trust literature. A test of the proposed theoretical framework was conducted in a field setting utilizing a matched design of both project managers and their clients. The results showed that reliable project performance positively impacted client loyalty intentions and service-oriented OCBs positively impacted client secondary retention. An interaction showed that in low cognitive trust situations the project manager's behavior of reliable project performance facilitated the obtainment of client loyalty intentions.

Pauland Mcdaniel (2004) examined the relationship between interpersonal trust and virtual collaborative relationship (VCR) performance. Findings from a study of 10 operational telemedicine projects in health-care delivery systems were presented. The results presented here confirmed, extended, and apparently contradicted prior studies of interpersonal trust. Four types of interpersonal trust–calculative, competence, relational, and integrated–were identified and operationalized as a single construct.

Support for an association between calculative, competence, and relational interpersonal trust and performance was found. Finding of a positive association between integrated interpersonal trust and performance not only yielded the strongest support for a relationship between trust and virtual collaborative relationship performance but also contradicted prior research. The findings indicated that the different types of trust are interrelated, in that positive assessments of all three types of trust are necessary, if virtual collaborative relationship is to have strongly positive performance. The study also established that if any one type of trust is negative, then it is very likely that virtual collaborative relationship performance will not be positive. Findings indicated that integrated types of interpersonal trust were interdependent, and the various patterns of interaction among them were such that they were mutually reinforcing.

Serva, Fuller, and Mayer (2005) developed and investigated the concept of reciprocal trust between interacting teams. Reciprocal trust was defined as the trust that results when a party observes the actions of another and reconsiders one's trust-related attitudes and subsequent behaviors based on those observations. Twenty-four teams of systems analysis and design students were involved in a 6-week controlled field study, focused on the development of an information systems project. Each team was responsible for both developing a system (development role) and for supervising the development of a system by another team (management role). Risk-taking actions exhibited by one team in an interacting pair were found to predict the other team's trustworthiness perceptions and subsequent trust. The level of trust formed in turn predicted the team's subsequent risk-taking behaviors with respect to the other team. This pattern of reciprocal trust repeated itself as the teams continued to interact over the duration of the project, thus supporting our model of reciprocal trust. Findings also indicated that trust and trust formation can occur at the team level.

The rapid growth of personal email communication, instant messaging, and online communities has brought attention to the important role of interpersonal trust in online communication. An empirical study was conducted focusing on the effect of empathy on online interpersonal trust in textual IM. To be more specific, the relationship between empathic accuracy, response type, and online interpersonal trust was investigated. The result suggested both empathic accuracy and response type have significant influence on online interpersonal trust. The interaction between empathic accuracy and response type also significantly influenced online trust. Interestingly, the results imply a relationship between daily trust attitude and online interpersonal trust. People who were more trusting in their daily life may experience more difficulty in developing trust online. There was also some evidence to suggest that different communication scenarios may had an influence on online trust (Feng, Lazar, & Preece, 2007).

Vlaar, Vanden, and Volberda (2007) suggested that the degrees to which managers trust and distrust their partners during initial stages of cooperation leave strong imprints on the development of these relationships in later stages of collaboration. This derives from the impact of trust and distrust on (a) formal coordination and control, (b) interorganizational performance, and (c) the interpretations that managers attribute to the behavior of their partners. Collectively, the authors' arguments give rise to a conceptual framework that indicates that there is a high propensity for interorganizational relationships to develop along vicious or virtuous cycles. By integrating and reconciling previous work on the trust-control nexus and by emphasizing the dynamics associated with it, it was contributed to a more comprehensive and refined understanding of the evolution of interorganizational cooperation.

Trusting relationships are increasingly considered vital for making teams productive. It was proposed that cooperative management of conflict can help team members to be convinced that their teammates are trustworthy. Results from 102 organizations in China supported the theorizing that how teams to manage conflict with each other affects within-team conflict management. Specifically, cooperative conflict between teams helped teams to manage their internal conflicts cooperatively that strengthens trust that in turn facilitates team performance. Results provided support for managing conflict cooperatively as a foundation for trusting, productive relationships in China as well as in the West (Hempel, Zhang, & Tjosvold, 2009).

Secure attachment is a healthy attachment style that enables individuals to work autonomously as well as with others when appropriate. Secure attachments are characterized by internal regulatory mechanisms that allow individuals to be flexible and constructive in their interpersonal relationships given in the model, which incorporated hope, trust in one's supervisor, and burnout as explanatory variables that translated the benefits of secure attachment into better supervisor-rated task performance. Among 161 employees of an assisted living center and their supervisors, secure attachment had a significant, positive relationship with hope, trust, and burnout, but only trust had a significant, positive relationship with supervisor-rated performance. These results indicated that secure attachment should be considered a positive psychological strength that has important implications for working adults (Simmons, Gooty, Nelson, & Little, 2009).

Given the growing importance and complexities of telework and the challenges associated with knowledge sharing, teleworkers and their propensity to share knowledge was investigated by investigating if the relational qualities of teleworkers in the form of trust, interpersonal bond, and commitment act to impact teleworker knowledge sharing and also investigated how telework's altered spatial and technical interactions shape knowledge sharing by testing the contingent role of technology support, face-to-face interactions, and electronic tool use. Results using matched data from 226 teleworkers supported the role of teleworker trust, interpersonal bond, and commitment in predicting knowledge sharing. Moreover, the impact of trust on knowledge sharing was found to be moderated by technology support, face-to-face interactions, and use of electronic tools, whereas the impact of commitment is contingent upon the use of electronic tools (Golden & Raghuram, 2010).

8.5 JOB INVOLVEMENT

Blau and Boal (1987) predicted that various combinations of organizational commitment and job involvement will have distinct consequences for organizations. For example, employees who exhibit both high organizational commitment and high job involvement should be the least likely to leave the organization. Employees with low levels of organizational commitment and job involvement (apathetic) should be the most likely to leave the organization voluntarily. Finally, Blan and Boal designated employees with high job involvement and low organizational commitment lone wolves and called employees with low job involvement and high organizational commitment corporate citizens. Because of their stronger organizational identification, corporate citizens were predicted to leave the organization less frequently than lone wolves.

The construct of job involvement is somewhat similar to organizational commitment in that they are both concerned with an employee's identification with the work experience. However the constructs differ, in that job involvement is more closely associated with identification with one's immediate work activities, whereas organizational commitment refers to one's attachment to the organization (Brown, 1996). It is possible, for example, to be very involved in a specific job but not be committed to the organization or vice versa (Blau & Boal, 1987).

Blau and Boal (1987) predicted that various combinations of organizational commitment and job involvement will have distinct consequences for organizations. For example, employees who exhibit both high organizational commitment and high job involvement should be the least likely to leave the organization. Employees with low levels of organizational commitment and job involvement (apathetic) should be the most likely to leave the organization voluntarily. Finally, Blau and Boal designated employees with high job involvement and low organizational commitment lone wolves and called employees with low job involvement and high organizational commitment corporate citizens. Because of their stronger organizational identification, corporate citizens were predicted to leave the organization less frequently than lone wolves.

Gable, Myron, and Frank (1994) studied and addressed the effect of job involvement on the relationship between Machiavellianism and job performance. The results showed a significant effect for managers who perceive themselves as possessing high levels of job involvement. No effect was found for managers who perceived themselves as having low levels of job involvement.

Using a measure created by Paullay et al. (1994) to differentiate job involvement from work centrality, Diefendorff et al. (2002) found a small but significant correlation ($r = 0.19$, $P < 0.05$) between job involvement and supervisor–rated in-role performance. Finally, Rotenberry and Moberg (2007) using the same measure of job involvement as Diefendorff et al. (2002) reported a small but significant positive correlation ($r = 0.15$, $P < 0.05$) between job involvement and in-role performance.

Brown and Leigh (1996) investigated the process by which employee perceptions of the organizational environment are related to job involvement, effort, and performance. The researchers developed an operational definition of a psychological climate to their own well-being. Perceived psychological climate was then related to job involvement, effort, and performance in path-analytic framework. Results showed that perceptions of a motivation and involving psychological climate were related to job involvement, which in turn was related to effort.

Research studies over the past two decades, which have explored the construct of job involvement, have approached it from two different perspectives (Sekeran, 1989; Sekeran & Mowday, 1981). First when viewed as an individual difference variable, job involvement is believed to occur when the possession of certain needs, values, or personal character-istics predispose individuals to become more or less involved in their jobs. For instance, Rabinowitz and Hall (1977) in their review of literature on job involvement found that individual characteristics such as age, education, sex, tenure, need strength, level of control, and values were linked to job involvement. The second perspective views job involvement as a response to specific work situation characteristics. In other words, certain types of jobs or characteristics of the work situation influence the degree to which an individual becomes involved in his/her job. For example, research has demonstrated that job involvement has been related to job characteristics such as task autonomy, task significance, task identity, skill variety, and feedback and supervisory behaviors, such as leader consideration, partici-pative decision-making, and amount of communication (Brown, 1996).

In a study in New Zealand, Guthrie (2001) showed a positive asso-ciation between use of high-involvement work practices (HIWPs) and employee retention and firm productivity.

Cohen (1999) research supported the important status of job involve-ment as an antecedent to organizational commitment. Specifically, Cohen argued that those individuals with high levels of job involvement, which stems from positive experiences on the job (Kanungo, 1979;), make attri-butions from these experiences to the organization.

Ahman and Ansari (2000) conducted a study on craftsman from various small-scale industries and noted that job involvement was influenced by the interaction between income and job tenure.

Srivastava (2001) conducted a study to examine job involvement and mental health among 60 executives and 15 supervisors with work experience ranging from 8 to 30 years, result revealed that executives felt more involved in the job than the supervisors. There was a signifi-cant association between job involvement and mental health.

Lassak et al. (2001) argued that occupation specific measures of job involvement should be created and consequently developed a measure of "salesperson job involvement." Their study uncovered a significant but positive relationship between one facet of their measure, "relationship" involvement and performance.

It was found that when involvement is low, the relationship one has with the company cannot be really strong. On the other hand, when involvement is high, the relationship stays strong even at low level of satisfaction (Ashok, 2002).

Allam (2007) examined job involvement of bank employees in relation to job anxiety, personality characteristics, job burnout, age, and tender. The result revealed that the job anxiety, job burnout, age, and tender were significant related to job involvement.

Fostering job involvement is an important organizational objective because many researchers consider it to be a primary determinant of organizational effectiveness (Pfeffer, 1994) and individual motivation (Hackman & Lawler, 1971). These links stem from the theoretical notion that being immersed in one's work increases motivational processes, which in turn influence job performance and other relevant outcomes like turnover and absenteeism (Diefendorff et al., 2002).

Mishra and Wagh (2004) conducted a study on public and private sector executive on job involvement dimension. Two groups of executives differ significant on mean score. Further they pointed out that reward. Work culture and environment, challenging job delegation of authority and responsibility were found to be potential factors for job involvement.

Usually, the higher the one's identification or involvement with a job, the greater is the job satisfaction (Schultz & Schultz, 2004).

Job involvement is the psychological identification with one's job. Recent trends in sales organizations have heightened the need for increased job involvement among salespeople. Little research was done to investigate the relationship of job involvement to demographic, job situational, and market variables in a sales setting. Results of a survey of 417 field salespeople revealed support for associations between job involvement and these variables. (Marshall, Greg, Lassak, & Moncrief, 2004).

Employee involvement is an organizational phenomenon that has received increasing empirical attention. Although much research has examined the outcomes of involvement at the organization level, arguments can be made for exploring involvement at the work-unit level and for investigating the processes by which a unit-level climate of involvement may be created or emerge. Building on largely untested suggestions that such processes are likely to be motivational and initiated by employees' immediate supervisors, two concepts of managerial perceptions and leadership were incorporated into a work-unit level model of involvement climate. In particular, the indirect association of managerial perceptions about subordinates' ability to perform and about the utility of organizational practices for facilitating performance, as well as the direct association of transformational leadership, with a climate of involvement was examined. The association of involvement climate with citizenship, absenteeism, and voluntary turnover was also considered. Using structural equation modeling in a sample of 167 work units, results indicated that leadership fully mediates the relationship between managers' perceptions about their subordinates and climate. Further, climate partially mediated and fully mediated the relationship between leadership and citizenship, and absenteeism, respectively (Richardson & Vandenberg, 2005).

Mishra and Shyam (2005) conducted a study to find out the relationship of social support and job involvement in prison officers. Job involvement scale developed by Kapoor and Singh and serial support scale developed by Cohen et al. were administered on a sample of 200 prison officers. The results showed that serial support and belonging support have significant positive relationship with job involvement. Stepwise multiple regression analysis suggests that overall serial support is a significant predictor of job involvement his prison officers. The other predictors are belonging support, appraisal support, and tangible support.

Allam (2007) conducted a study on bank employees and observed that personal accomplishment, one of the facets of job burnout was found significant related to job involvement among the bank managers.

The relationships between age, aspects of tenure, locus of control, job involvement, and boundary spanning behavior (BSB) were examined using path analysis for 281 scientists and engineers. It was found that locus of control and age were significant determinants of job involvement. It was also shown that loci of control and job involvement were significant determinants of boundary spanning behavior (Dailey & Morgan, 1978).

Emery and Barker (2007) suggested that the organizational commitment of customer contact personnel was significantly correlated with customer satisfaction but not profit and productivity. On the other hand, the job involvement of customer contact personnel was significantly correlated with customer satisfaction, profit, and productivity. There was a significant difference between the team and nonteam structures for job involvement, but not for the organizational commitment of customer contact personnel.

Dimitriades (2007) explored the usefulness to highlight the nature of interrelationship(s) between service climate and job involvement in impacting customer-focused organizational citizenship behaviors (OCB) of frontline employees in a diverse cultural context, at the crossroads of East and West. He provides empirical evidence of the applicability in Greek service contexts and illuminates the complex nature of interrelationships between organizational climate for service and job involvement in predicting customer-oriented organizational citizenship behaviors, expanding the OCB literature.

Teamwork goal orientation is positively related to intrinsic motivation and job involvement (Zuang, 2008).

Ouyang (2009) conducted a study to explore the caused relationship among the job uncertainty, job involvement, job stress, and job performance of banking service personal under the economic depression. The empirical result of study found that job instability of banking service personnel has negative influence on job performance and job involvement. Furthermore, job instability had a significant positive influence on job stress; job stress has a positive influence on job involvement and job performance. The mediating effect of job stress and job involvement positively influences job performance, the result further revealed that the most important factor in job performance is job involvement and second factor is job stress.

Uygur and Gonca (2009) studied the level of organizational commitment and the job involvement of the personnel at Central Organization of Ministry of Health in Turkey. In total, 210 subjects, selected randomly, were distributed the questionnaire forms. Of the questionnaires, 180 of them (86%) returned and 168 of them were regarded valid and acceptable and analyzed. A moderate positive correlation was found out between organizational commitment and job involvement ($r = 0.44$). In the light of this, there is a significant correlation between organizational commitment and job involvement, though not very strong.

Past studies that have examined correctional staff support for rehabilitation and punishment policies have produced conflicting results. Most studies have focused on personal characteristics, including age, gender, job position, tenure, education, marital status, prior military service, and race. To expand the area of inquiry and assess the potential antecedents of the work environment regarding correctional staff support for inmate rehabilitation or punishment the impact of job stress, job involvement, and organizational commitment on staff attitudes toward the rehabilitation or punishment of inmates was examined. Findings indicated that job involvement and organizational commitment positively influenced correctional staff support for rehabilitation policies; however, job stress and job satisfaction did not have a significant effect on correctional staff attitudes toward either punishment or rehabilitation (Lambert, Hogan, & Oko, 2009).

If there is serious "financial tsunami," the financial service personnel may lose confidence and this may influence their job involvement, and even, lose their job. It was attempted to understand the critical factors of their job involvement after suffering from the "financial tsunami" attack. Some suggestions were put forward to restore the job involvement of the financial service personnel. The empirically discovered "perceived orga-nizational support" (POS) has the significant direct effect on job involve-ment of employees. Peer relationship positively had an influence on job involvement through perceived organizational support; the direct effect was weaker than the indirect effect. These results demonstrated that peer relationship was conducive to job involvement of the financial service personnel via the five constructs of perceived organizational support. In another result, the direct effect of financial service personnel's guanxi networks on job involvement was weaker than the indirect effect on job involvement (Hao, Jung, & Yenhui, 2009).

Research indicated that job burnout is a negative response that is harmful to the employee and to the organization. Depersonalization, emotional exhaustion, and feeling a lack of accomplishment at work are all dimensions of job burnout. The association of job involvement, job stress, job satisfaction, and organizational commitment with burnout among correctional staff was examined. The findings highlighted the significance of these variables in relation to burnout. Specifically, job satisfaction had an inverse relationship with emotional exhaustion, deper-sonalization, and a sense of reduced accomplishment at work, whereas job stress had a significant positive relationship with depersonalization and emotional exhaustion. Job involvement also had a positive associa-tion with emotional exhaustion, whereas commitment to the organization had no relationship with any of the three dimensions of burnout (Griffin, Lambert, Tucker-Gail, & Baker, 2009).

Job involvement is a principal factor in the lives of most people; employees in the workplace are mentally and emotionally influenced by their degree of involvement in work. Using the data from the National

Administrative Studies Project III, Word and Park (2009) studied empirically and compared the level of job involvement between managers in the public and nonprofit sectors and explores different aspects, including demographic, managerial, and institutional factors that contribute to the apparent differences. The results of the study indicated that the mean level of nonprofit managers' job involvement was significantly greater than for public managers. Each sector had specific variables that significantly and uniquely contributed to job involvement. Overall, the results suggested a need to more fully investigate the various mechanisms and functions of situational and organizational contexts, organizational norms, and culture that were associated with job involvement regardless of sector.

Individuals have been described as job involved if they view it as important to their life interest (Dubin, 1956) and perceive performance as central to their self-esteem (Gurin et al., 1960). Vroom (1962) describes a person as ego involved in a job by the level of his self-esteem which is affected by his perceived level of performance. Other conceptual way of describing job involvement is the "degree to which a person is identified psychologically with his work" or "the importance of work in his total self-image" (Lodahl & Kejner, 1965). Such a psychological identification with work may result partly from early socialization training during which the individual may internalize the value of goodness of work. Lodahl and Kejner (1965) emphasized that during the process of socialization, certain work values are injected into the individual that remains even at the later stage in the form of attitude toward job (Azeem, 2010).

The study examined the influence of personality hardiness, job involvement and job burnout among teachers from one of the central universities in India. The findings of the study revealed that personality hardiness and job involvement resulted from job burnout of teachers. Job involvement was found to be negatively and significantly correlated with depersonalization and positively with personal accomplishment dimensions of burnout. Commitment, challenge, control, and total personality hardiness were found to be negatively related with emotional exhaustion. Commitment and total personality hardiness were also found to be negatively related with depersonalization. Stepwise multiple regression analyses summarize that commitment and total personality hardiness were the predictors of burnout among teachers (Azeem, 2010).

KEYWORDS

- organizational climate
- globally distributed team
- structural equation modeling
- organizational citizenship behavior
- self-efficacy

CHAPTER 9

RESEARCH DESIGN

CONTENTS

9.1 RATIONALE OF THE STUDY

Globalization, information technology has eased the business and at the same time increased competitive demands for the organizations, which forced many organizations to increase levels of flexibility and adaptability in their operations. The need to become more service oriented is making organizations challenge conventional structures and theories of behavior within organizations. A large number of such organizations are exploring the virtual environment as one means of achieving increased responsiveness globally. In particular, the use of virtual teams appears to be on the increase not in west but in Asian continent also. However, the increased use of virtual teams has not been accompanied by contemporaneous research efforts to understand better the psycho-social contributors to effective virtual teams.

Virtual teams are creating new challenges for business leaders. To work in diverse culture with geographical differences and time differences, leadership becomes challenging to the leader and makes difficult to follow structure, judge or evaluate the performance, or inspire them and enable team members/workers to identify with the organization.

Most large organizations have multiple divisions and locations. Smaller companies may have team members working at different sites. The current trend toward working collaboratively with suppliers, partner organizations, and clients requires diversely located team members effectively together. Virtual teams have many obstacles to overcome. They have to deal with the usual project challenges, and in addition may face very significant communication, motivational and organizational issues.

According to Avolio, Kahai, and Dodge (2001), there are new frontiers rapidly opening, focusing on what constitutes effective leadership in the information environment. It is strongly believed that both the research community and organizations can benefit by examining the topic of leadership in virtual teams. Researchers and practitioners need to know the role of a virtual team's "virtuality" in influencing leader–follower interactions and its effects on leadership.

As the technological infrastructure necessary to support the virtual teams is now readily available, further research on the range of issues surrounding virtual teams is required if we are to learn how to manage them effectively. While the findings of team research in the traditional environment may provide useful pointers, the idiosyncratic structural and contextual issues surrounding virtual teams call for specific research attention.

As trust is an integral part of any organization, that is, whether between employees themselves or between management and employees and in a scenario where there is no face-to-face interaction, how would trust develop? Again within the reach of superiors and that too 24 × 7, due to information technology, how a virtual worker is involved in his job? How much in virtual atmosphere a worker's belief about his capabilities to produce desired performance will affect his life? And yes, besides leadership style, job involvement, trust, self-efficacy and, how will be the whole organizational climate without physical environment. Keeping the above scenario and factors in mind the researchers decided to study this challenging work scenario, trend of future organizations in today's world of globalization and information technology, and took initiative to explore organizational climate in relation to leadership, self-efficacy, trust, job involvement in virtual workers.

9.2   OBJECTIVES

1. To study whether perceived organizational climate is related to leadership effectiveness among the virtual workers in different organizations.
2. To examine the relationship of perceived organizational climate with self-efficacy among virtual workers in different organizations.
3. To find out whether perceived organizational climate is related to trust among virtual workers in different organizations.
4. To find out whether perceived organizational climate is related to job involvement among virtual workers in different organizations.
5. To find out the intercorrelation among the different criterion variables, that is, leadership effectiveness, self-efficacy, interpersonal trust, and job involvement.
6. To find out the best predictors of leadership effectiveness, self-efficacy, interpersonal trust, and job involvement with regard to perceived organization climate and its domains.

9.3   HYPOTHESES

The study intended to test the following hypotheses:

1. There will be significant positive relationship between perceived organizational climate and leadership effectiveness among virtual workers of different organizations.
2. There will be significant positive relationship between perceived organizational climate and self-efficacy among virtual workers of different organizations.
3. There will be significant positive relationship between perceived organizational climate and trust among virtual workers of different organizations.
4. There will be significant positive relationship between perceived organizational climate and job involvement among virtual workers of different organizations.
5. There will be positive relationship between the different criterion variables under the study.
6. There will be significant positive relationship between different criterion variables under study among the virtual workers of different organizations.

9.4   VARIABLES

- Predictor variables: Perceived organizational climate and its domains–performance standards, communication flow, reward system, responsibility, conflict resolution, organizational structure, motivational level, decision-making process, support system, warmth, identity problems
- Criterion variables

    i. Leadership effectiveness and its domains–interpersonal relations, intellectual operations, behavioral and emotional stability, ethical and moral strength, adequacy of communication, operations as a citizen
    ii. Self-efficacy

iii. Trust and its domains–trust in supervisor and trust in organization

iv. Job involvement

9.5 RESEARCH DESIGN

Fisher (1951) remarked, "If the design of an experiment is faulty, any method of interpretation, which makes it out to be decisive, must be faulty too." Pointing out the importance, Kerlinger (1973) remarked that the chances of arriving at accurate and valid conclusions are better with social designs than with unsocial ones. If the design is faulty, one can come to no clear conclusions. For the study, correlation design and multiple regression analysis were used.

9.5.1 CORRELATIONAL DESIGN

The important defining features of correlational research are that the researchers do not directly manipulate the variable under study. As a general statement, we may say that research is likely to be correlational in nature whenever all the variables under study concern properties of the subject which are either inherent to the subject (e.g., age, sex, intelligence, etc.) or the result wholly or in part of prolonged experienced (e.g., aspects of personality).The coefficient of correlation tells us the way in which two variables are related to each other. How the change in one is influenced by the change in other may be explained in terms of direction and magnitude of these measures.

The investigation dealt with the interrelationship between perceived organizational climate, leadership effectiveness, interpersonal trust, self-efficacy, and job involvement among virtual workers in different organizations. So the investigator thought it proper to adopt correlation design.

The correlational coefficient in the variables was calculated in the investigation with equal appropriateness.

As per the hypotheses, the study was regarding the interactive effects of the variables. Correlational design is most appropriate measure to examine the relation among the variables. Thus, in the study the following design is used.

CORRELATIONAL DESIGN
Order of administration of measurement devices: Randomly

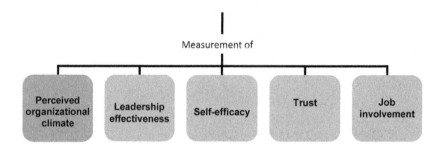

KEYWORDS

- job involvement
- decision-making
- globalization
- information technology
- self-efficacy

CHAPTER 10

METHODOLOGY

CONTENTS

10.1 RATIONALE OF THE SAMPLE

In the past decade, the fast-changing pace of technology has made a huge impact on the business world. The radical change in market took place due to liberalization, globalization, and technological advances. The result of that customer can get a product made in America or anywhere in the world, and to survive in local market, businesses are expanding. India's growth story shows that those who embraced change post 1991 have not only survived but also excelled. With globalization, organizations become more service oriented with the help of fast technology, which has meta-morphosed the work culture and work setting. If a customer in Jaipur wishes to buy an organizational book from America or New Delhi, he/she does not need to go to America or New Delhi, not even to a bookshop in Jaipur. Yes, he/she can receive it within 24 h at home itself. How is it possible? Globalization and technology have changed the work culture and so the workforce. As a result, globalization has become a big challenge for organizational behavior experts.

Organizational behaviorists have helped to provide new alternative work arrangements by the organizations to today's workforce. This workforce is very different and sounds interesting. Here, the individual is working literally "from the road" and from place to place or from customer to customer by car or airplane. In all cases, the worker remains linked electronically with the home office.

The number of workers who are telecommuting is growing daily, with organizations such as Cisco Systems reporting that more than 50% of their workers telecommute at least part of the time. Many organizations such as IBM embrace virtual work not only as a means of helping employees with work–life balance issues but also for bottom-line interests.

For multinational companies and organizations that work across the borders, cost cutting is one of the issues with increasing their efficiencies in terms of customer focus. This might be proving profitable as in case of IBM which reports that firm estimates $100+ million is saved each year.

But as organization behaviorist and psychologist whether really these virtual working individuals are able to create balance in their work life and is increasing workforce feels satisfied from this kind of working is a focus area to study in coming future. In today's organizations, leadership is a very important aspect as teams' performance and organizations' output are dependent on it. How easy or difficult it would be when no face-to-face traditional leaders are there in vertical hierarchy or organizational structure? Whether the people who are becoming virtual workers have belief in them to excel and perform in unstructured organizations? Whether involvement in work would be more at home or at traditional offices and what trusting relationships would be formed in office setting due to formal and informal meetings including gossips?

As this concept is new and little is known about it scientifically, in the past, not many relevant researches in psychology have been taken to throw light on virtual workforce; the researcher thought it is relevant to study this new area as it is becoming one of the challenges in the field of organizational behavior in the present era of digital age. With these questions in mind, researcher took virtual workers as sample of this study.

10.2 DISTRIBUTION OF SAMPLE

In the study, 100 virtual workers were selected as subjects. The sample of virtual workers was selected adopting purposive sampling method based on the availability as well as on the willingness of the workers to participate in this study. Respondents were selected from various organizations in IT, ITES, and hospitality sectors residing in Delhi and National Capital Region (NCR). The criteria for selection of subjects were as follows (Table 1):

TABLE 1 Criteria For Subjects Selection

S. No.	Criteria	Description
1.	Age	25–35 years
2.	Job experience	At least 3 years
3.	Education	Postgraduation or equivalent
4.	Level of employee	Executive level
5.	Gender	Only males
6.	Organization type	IT, ITES, hospitality
7.	Marital status	Married

10.3   MEASURES AND SCORING METHODS

10.3.1   ORGANIZATIONAL CLIMATE INVENTORY–CHATTOPADHYAY & AGARWAL (1976)

10.3.2   LEADERSHIP EFFECTIVENESS SCALE–TAJ HASEEN (2001)

10.3.3   GENERALIZED PERCEIVED SELF-EFFICACY SCALE–SCHWARZER & BORM (1997)

10.3.3   ORGANIZATIONAL TRUST INVENTORY–NYHAN AND MARRLOWE (1997)

10.3.4   JOB INVOLVEMENT–SINGH (1984)

10.3.1   ORGANIZATIONAL CLIMATE INVENTORY–CHATTOPADHAY & AGARWAL (1976)

10.3.1.1   DESCRIPTION OF INSTRUMENT DEVELOPMENT; ORGANIZATION CLIMATE INVENTORY (FORM A)

The process of instrument development started with Form A of OCI. Two senior members of the research team scanned through the literature and formulated a list of 20 dimensions of organizational climate. Likert scale was used for item construction.

With 116 items, OCI Form A was too lengthy to be used effectively with a population of busy hospital personnel. Therefore, it was necessary to reduce its length and drop some of its items. In order to achieve this objective, Form B was developed using two methods: (1) application of the OCI Form A on hospital personnel and factor analysis of the data; rejection of items with low factor loadings and (2) expert ratings for the suitability of each item of OCI Form A.

10.3.1.2   ORGANIZATIONAL CLIMATE INVENTORY FORM B: EXPERT RATINGS

OCI Form A was sent out to 85 experts in organizational behavior for their ratings. These persons were actively engaged either in consultancy, research, or teaching in the field of organizational behavior/management. Of these, 25 experts sent their ratings. Among these respondents, several were eminent persons in the profession. Some of the experts offered valuable suggestions which were incorporated while constructing OCI Form B.

10.3.1.3 DATA ANALYSIS

Data were collected from a large hospital using OCI Form A. A stratified random sample was drawn and responses of 1200 respondents on 116 items were collected. These responses were calculated using summated scoring method, and intercorrelations between 116 variables were calculated. As many as 66,670 coefficients of correlations were carried out. This huge matrix of intercorrelations was factor analyzed. Thirty three factors were extracted. Items were selected using judge's ratings and factor analysis data from hospital staff. All items with ratings above median judge's ratings and high factor loading were selected. In all, 70 items were selected.

10.3.1.4 DIMENSIONS OF OCI

Incorporating judge's ratings and using insight gained from factorial data, the 70 items thus selected formed independent 11 scales.

S. No.	Dimensions	Item Nos.
1.	Performance standards	6, 9, 10, 13, 30, 31, 57
2.	Communication flow	12, 17, 24, 34, 37, 38, 49, 52, 61, 65, 67
3.	Reward system	29, 41, 54, 66
4.	Responsibility	4, 16, 27, 40
5.	Conflict resolution	1, 18, 23, 42, 44, 45, 46
6.	Organizational structure	14, 19, 21, 35, 47
7.	Motivational level	28, 32, 51, 56, 59, 68, 69
8.	Decision-making process	2, 15, 25, 36, 43, 62, 70
9.	Support system	3, 5, 7, 8, 20, 48, 53, 55, 58
10.	Warmth	26, 39, 60, 63,64
11.	Identity problems	11, 22, 33, 50
	Total	**70 items**

10.3.1.5   VALIDITY OF OCI FORM B

A) FACTORIAL VALIDITY
Items for OCI Form B were selected using factor loadings on OCI Form A factor analysis as described above. Factor analysis being one of the criteria for item selection amounts to selecting valid items.

B) FACE VALIDITY
As described above, judge's ratings were used for the selection of OCI Form B items.

C) ITEM VALIDITY
For computing item validity, correlations with total organization climate scores were computed. These correlations show that all items had highly significant correlation with total OC score; P being .001 for 68 out of 72 items.

10.3.1.6   RELIABILITY OF OCI FORM B

Reliability coefficient by Spearman–Brown formula was .898, which shows that there was high internal consistency in the instrument, and hence it was highly reliable.

10.3.2   LEADERSHIP EFFECTIVENESS SCALE, TAJ HASEEN (2001)

After going through the related literature and tools available to assess the leadership behavior/effectiveness, and consultations with experts in the field, the researcher initially pooled 134 statements distributed over six operationally defined areas of the scale. Later these statements were given to 30 experts to evaluate the content accuracy, coverage, conceptualization, etc. Based on 85% unanimity (agreement) of the judges, only 90 items were retained for the try-out form of the scale.

10.3.2.1   TRY-OUT OF THE LES

The 90 statements were arranged at random and were administered on a sample of 435 group members from educational institutions, industries, and business organizations representing government, quasi-government, and private management of Bangalore city. The sample was selected using stratified proportionate random sampling technique.

10.3.2.2   SCORING OF THE STATEMENTS

The respondents were asked to indicate the behavior of their group leader or head on the five points given against each statement. All the statements were scored giving a weightage to each of the alternative responses of the statement in the pattern given below for all the positive items:

Always (A)	5
Often (B)	4
Occasionally (C)	3
Rarely (D)	2
Never (E)	1

For all the negative items of the scale, scoring was reversed ranging from 1 to 5. Items marked with black dot are negative items. The total score of a respondent ranges from 90 to 450 in the try-out form of the scale, and subsequently the total score of a respondent ranges from 79 to 395 in the final scale as 11 items were eliminated.

TABLE 2   Total Number of Items Distributed Over Different Areas in the Final Scale After Item Analysis

S. No.	Areas	Serial number of items in the scale	Total number of items in each area
I	Interpersonal relations	1, 7, 13, 15, 16, 22, 23, 33, 34, 38, 47, 53, 57, 58, 67, 73	16
II	Intellectual operations	3, 5, 12, 17, 18, 39, 40, 41, 51, 56, 59, 60, 75	13
III	Behavioral and emotional stability	10, 11, 48, 55, 61, 68, 69, 72, 74, 76, 77	11
IV	Ethical and moral strength	6, 8, 9, 20, 21, 26, 27, 28, 29, 30, 31, 35, 37, 42, 50, 66, 70, 71, 79	19
V	Adequacy of communication	2, 4, 14, 32, 45, 46, 49, 54, 62, 64, 65	11
VI	Operations as a citizen	19, 24, 25, 36, 43, 44, 52, 63, 78	9
	Total		79

' *10.3.2.3   FINAL SCALE*

The *t*-test was computed for all the 90 statements. Based on the *t* values, only those statements were selected for the final form of the scale which were significant at .05 level or higher. Only 79 items got selected and retained in the final scale, eliminating 11 items which were not significant even at .05 level. Table 2 shows the final scale items after item analysis with total number of items with their serial numbers distributed over different areas. Table 3 shows the distribution of positive and negative items in Leadership Effectiveness Scale (LES) in final scale.

TABLE 3   Distribution of Positive and Negative Items in Les in Final Scale

S. No.	Items	Serial numbers as given in final scale	Total number of items	%
1.	Positive	1, 2, 3, 4, 5, 7, 8, 10, 12, 13, 14, 15, 16, 17, 18, 19, 20, 21, 22, 23, 24, 25, 26, 27, 28, 30, 32, 35, 36, 38, 39, 42, 43, 44, 45, 46, 47, 48, 50, 51, 52, 53, 56, 57, 59, 60, 63, 64, 66, 67, 73, 74, 75, 78	55	69.62
2.	Negative	6, 9, 11, 29, 31, 33, 34, 37, 40, 41, 49, 54, 55, 58, 61, 62, 68, 69, 70, 71, 72, 76, 77, 79	24	30.38
	Total		79	100

10.3.2.4   RELIABILITY

Two types of reliabilities were established for the scale, that is, test–retest and split-half reliabilities, using a sample of 140 group members from educational, industrial, and other organizations representing different managements in the city of Bangalore.

The test–retest reliability coefficient was found to be .60 with a time gap of 2 weeks and split-half reliability coefficients by odd and even and first half and second half (1st half vs. 2nd half) method were found to be .64 and .67, respectively. After the application of Spearman–Brown prophecy formula to split-half reliability, coefficient® gave rise to coefficients of .78 and .80.

10.3.2.5   VALIDITY

i.   Content validity: The scale possesses content validity because the statements were selected based on 85% unanimity of experts on content adequacy, conceptualization, and distribution of statements over different areas.

ii.   Intrinsic validity: The scale had intrinsic validity, since the index of reliability was taken as the intrinsic validity by working out the square roots of the reliability coefficient of the scale. The test–retest reliability was .60 and split-half reliability was .78 and its square roots were found to be .77 and .88, respectively using odd and even method. The split-half reliability of the scale using 1st half and 2nd half method was found to be .67 and rose to .80 with the computation of reliability index.

iii.   Item validity: The items for the final scale were taking the t values of the items which had a significance of .05 level or higher. This indicates the high item validity.

iv.   Cross-validity: The cross-validity of the scale was established by selecting a different sample other than the one which was selected for the try-out of the scale. This was done in order to avoid the chance error within that particular sample calculated, which would have increased the probability of high reliability.

v.   Criterion-related validity: Three types of criterion-related validities were established for the scale. They are: (a) the correlation between LES (self) and LES (others); (b) t value calculated with the scores of LES (self) administered on two contrasted groups perceived by themselves as effective or ineffective leaders; and (c) t value calculated with the scores of the LES (others) administrated on two contrasted groups identified by their group members as effective or ineffective leaders.

10.3.3   GENERALIZED PERCEIVED SELF-EFFICACY SCALE SCHWARZER & BORN (1997)

Jerusalem and Schwarzer originally developed the German version of this scale in 1981, first as a 20-item version and later as a reduced 10-item version. The scale consists of 10 items, and four responses/choices were provided for each item, that is, (1) not at all true, (2) hardly true, (3) almost true, and (4) very true. Typical items are, "thanks to my resourcefulness, I know how to handle unforeseen situations, and when I am confronted with a problem, I can usually find several solutions." It has been used in numerous research projects, where it typically yielded internal consistencies between alpha = .75 and .91. This scale is not only parsimonious and reliable, it has also proven valid in terms of convergent and discriminate validity.

10.3.4 *ORGANIZATIONAL TRUST INVENTORY NYHAN &* *MARLOWE 1997*

The Organizational Trust Inventory (OTI) was designed to reflect the assumed differentiation of systems and personal trust in Luhmann's theory. It is a 12-item scale, with 8 items measuring trust in supervisor and 4 items measuring trust in the organization as a whole. The scale is a 7-point Likert-type format. The trust in supervisor subscale was developed to represent personal trust, that is, it is the immediate supervisor who is the most critical mediator of organizational or environmental complexity.

The first step in the scale validation process was to administer the OTI to several small groups to assess the instrument's reliability, validity, and assumed factor structure. Four groups were chosen:

- Pretest group 1: It included 31 students in a technical supervision course. Thirty of the participants were male, all were first-line supervisors or higher and all were 30 years of age or older.
- Pretest group 2: It included 23 employees who were students in a technical training program. Twenty-two of the 23 persons were male, the age range was from 22 to 56, and educational levels ranged from eighth grade to college graduate.
- Pretest group 3: It included 52 employees of a public utilities company. All were male, blue-collar workers. Of the respondents, 36.5% had less than a high school diploma. The remainder had high school degrees or above.
- Pretest group 4: It included 27 employees from a public housing authority. Twenty-two percent were supervisors, 44% were professionals, and 35% were clerical personnel. Seventy-two percent of the respondents had less than 1 year of experience with the agency, 35% had 2–10 years of experience, and 37% had more than 10 years of experience

10.3.4.1 RELIABILITY AND VALIDTTY

Internal consistency tests showed that each of the study groups' coefficient alphas was very high (.9597, .9599, and .9495, respectively). Confirmatory factor analysis (CFA) was used to test the validity of the dichotomized scale. CFA takes an a priori research perspective to multivariate data analysis, that is, the pattern of interrelationships among the study constructs is specified and grounded in established theory, for example, Luhmann's two-dimensional theory of trust. Discriminant validity is supported by the ×2 difference test. Convergent validity was assessed using the data from all study groups.

It exhibited both internal homogeneity and consistency, temporal reliability, and discriminant and convergent validity.

10.3.5. JOB INVOLVEMENT SCALE: SINGH (1984)

The preliminary schedule of the scale consisted of 72 items out of which 20 items were finally selected for the scale on the basis of item analysis. The 20 items which constituted the scale related to the following areas:

1. Intrinsic motivation
2. Attachment to work
3. Fulfillment of organizational demand
4. Commitment for work
5. Internalization of organization goals
6. Organizational identification

The items of the scale were framed in such a way that they can be used for measuring the degree of involvement of all the subjects irrespective of the nature of their work, organizations, and machines and tools they use.

10.3.5.1 RELIABILITY AND VALIDITY

The reliability of the job involvement scale was computed by the Cronbach's alpha coefficient technique and was found to be .85. The scores on job involvement scale of Lodahl and Kejner (1965) were used as the validation criteria for the present scale. The coefficient of correlation between the scores on the two tests was found to be .93 ($N = 400$). The results of the factor analysis indicate that 89% of the items had significant loading on factor-I, indicating the unidimensionality of the scale.

Scoring: Out of 20 items constituting the scale, 13 items were true keyed and remaining 7 items were false keyed. The possible scores of each items ranged from one to four.

True-keyed items (13)	False-keyed items (7)
1, 2, 3, 4, 5, 6, 7, 8, 9, 11, 12, 15, 20	10, 13, 14, 16, 17, 18, 19

Since the scale consists of both true-keyed and false-keyed items, two different patterns of scoring have to be adopted. The scores ranged in ascending order for the false-keyed items and in descending order for the true-keyed items. The following table provides guidelines for the scoring.

True-keyed items	Response alternatives	False-keyed items
4	Strongly agree	1
3	Agree	2
2	Disagree	3
1	Strongly disagree	4

The job involvement scores will be determined by the arithmetic summation of the scores endorsed to all the 20 items. Thus, maximum possible job involvement score will be 80 and the minimum 20. The lower scores indicate less involvement in the job and the high scores indicate more involvement in the job.

10.4 PROCEDURE

Virtual workers of different organizations were included as the sample of the study. To begin with, the information about the sample was collected initially from the company's website and social websites and then by meeting them face to face. Researcher built her rapport through social websites such as LinkedIn. After getting their informed consent and as per convenience of virtual workers, time and duration for filling questionnaires was fixed. Before administering the questionnaires, a rapport was established with the subjects and they were assured of the confidentiality of their response. First of all, they were asked to fill up a pro forma regarding general information (i.e., age experience, industry marital status, etc.). Out of this information, 100 subjects were selected according to the purpose. Then the questionnaire/tools were distributed among all subjects. All the questionnaires were administered as per instructions. Special care was taken in administering the test properly, minimizing the effect of fatigue by administering the test in three sessions. For assessing overview of virtual workers, a detailed interview for 10 virtual workers as per their availability was taken by the researcher. Each virtual worker's half an hour was taken for interview

10.5   STATITICAL ANALYSIS

1. Product movement correlation was applied to find out the relation between variables under study, that is, perceived organizational climate, leadership effectiveness, self-efficacy, trust, and job involvement.
2. Multiple regression was applied to find out the variance of effect of independent variables, that is, perceived organizational climate and its domains on dependent variable, that is, leadership effectiveness, self-efficacy, trust, and job involvement. Multiple regression analysis is method for studying the effects and magnitude of the effects of more than one independent variable on one dependent variable using principle of correlation and regression. It helps in studying complex interrelation between independent or predictor variable and a dependent or criterion variable.

KEYWORDS

- reliability
- validity
- Leadership Effectiveness Scale
- globalization
- cross-validity

CHAPTER 11

RESULTS

CONTENTS

As per feasibility, results are presented in two major parts.

11.1 PART I

The first aim of the study was to find out the correlation between perceived organizational climate and other variables in the study, that is, leadership effectiveness, self-efficacy, interpersonal trust, and job involvement among virtual workers.

A close analysis of the results for all the subjects gives us a picture of relationship between different variables under study. Table 1 shows that in total sample of virtual workers, perceived organizational climate and its dimensions performance standards, communication flow, reward system, responsibility, conflict resolution, organizational structure, motivational level, decision-making process, support system, warmth, and identity problems have been found to be significantly positively correlated with leadership effectiveness ($P < .01$) and with all its dimensions such as interpersonal relations ($P < .01$), intellectual operations ($P < .01$), behavioral and emotional stability ($P < .01$), ethical and moral strength ($P < .01$), adequacy of communication ($P < .01$), and operations as a citizen ($P < .01$). Factors related to perceived organizational climate and performance standards are found to be significantly correlated with all factors affecting leadership, interpersonal relations ($P < .01$), intellectual operations ($P < .01$), behavioral and emotional stability ($P < .01$), ethical moral strength ($P < .01$), and adequacy of communication ($P < .01$). Communication flow is the second domain of perceived organizational climate which is also found to be significantly correlated with interpersonal relations ($P < .01$), behavioral and emotional stability ($P < .01$), ethical moral strength ($P < .01$), and adequacy of communication ($P < .01$). Reward system is the third domain of perceived organizational climate and is found to be significantly correlated with all the factors of leadership effectiveness interpersonal relations ($P < .01$), intellectual operations ($P < .01$), behavioral and emotional stability ($P < .01$), ethical moral strength ($P < .01$), adequacy of communication ($P < .01$), and with leadership effectiveness in total. Responsibility is one of the domains of perceived organizational climate and is correlated with interpersonal relations ($P < .05$), behavioral and emotional stability ($P < .05$), and adequacy of communication ($P < .05$). Responsibility is found to be significantly correlated with ethical and moral strength

TABLE 1 Correlation Matrix of Total Sample of Virtual Workers

		Sup Trust	Org Trust	S Effic	IR	I Ops	B & ES	E & MS	A of Comm	Ops as Citi	Job Invol	Trust	L Effect	Per Std	Comm Rwd Sys Plo	Resp	Conf Reso	Org Strc	Moti Lvl	DM	SS	Warmth	Id Prob	OC
Sup Trust	Pearson correlation	1																						
	Sig. (2-tailed)																							
	N	100																						
Org Trust	Pearson correlation	-0.077	1																					
	Sig. (2-tailed)	.449																						
	N	100	100																					
S Effic	Pearson correlation	.055	-.103	1																				
	Sig. (2-tailed)	.585	.308																					
	N	100	100	100																				
IR	Pearson correlation	.224(*)	.141	-.247(*)	1																			
	Sig. (2-tailed)	.025	.162	.013																				
	N	100	100	100	100																			
I Ops	Pearson correlation	.134	.267(**)	-.137	.523(**)	1																		
	Sig. (2-tailed)	.182	.007	.175	.000																			
	N	100	100	100	100	100																		
B & ES	Pearson correlation	.138	.146	-.211(*)	.324(**)	.338(**)	1																	
	Sig. (2-tailed)	.170	.147	.035	.001	.001																		
	N	100	100	100	100	100	100																	
E & MS	Pearson correlation	.255(*)	.214(*)	-.303(**)	.660(**)	.533(**)	.517(**)	1																
	Sig. (2-tailed)	.010	.032	.002	.000	.000	.000																	
	N	100	100	100	100	100	100	100																
A of Comm	Pearson correlation	.260(**)	.181	-.224(*)	.531(**)	.400(**)	.549(**)	.593(**)	1															
	Sig. (2-tailed)	.009	.072	.025	.000	.000	.000	.000																
	N	100	100	100	100	100	100	100	100															
Ops as Citi	Pearson correlation	.157	.284(**)	-.062	.341(**)	.397(**)	.185	.185	.471(**)	1														
	Sig. (2-tailed)	.118	.004	.538	.001	.000	.066	.065	.000															
	N	100	100	100	100	100	100	100	100	100														
Job Invol	Pearson correlation	.087	.039	.042	.031	-.085	-.095	.087	.071	.075	1													
	Sig. (2-tailed)	.391	.703	.680	.756	.399	.348	.387	.484	.460														
	N	100	100	100	100	100	100	100	100	100	100													
Trust	Pearson correlation	.844(**)	.470(**)	-.007	.274(**)	.263(**)	.201(*)	.341(**)	.328(**)	.292(**)	.098	1												
	Sig. (2-tailed)	.000	.000	.948	.006	.008	.045	.001	.001	.003	.334													
	N	100	100	100	100	100	100	100	100	100	100	100												
L Effect	Pearson correlation	.270(**)	.273(**)	-.283(**)	.819(**)	.731(**)	.641(**)	.841(**)	.776(**)	.536(**)	.024	.386(**)	1											
	Sig. (2-tailed)	.007	.006	.004	.000	.000	.000	.000	.000	.000	.809	.000												
	N	100	100	100	100	100	100	100	100	100	100	100	100											

TABLE 1 Continued

		Sup Trust	Org Trust	S Effic	IR	1 Ops	B & ES	E & MS	A of Comm	Ops as Citi	Job Invol	Trust	L Effect	Per Std	Comm Flo	Rwd Sys	Resp	Conf Reso	Org Strc	Moti Lvl	DM	SS	Warmth	Id Prob	OC
Per Std	Pearson correlation	.080	.224(*)	-.176	.501(**)	.264(**)	.442(**)	.585(**)	.556(**)	.362(**)	.028	.192	.624(**)	1											
	Sig. (2-tailed)	.427	.025	.080	.000	.008	.000	.000	.000	.000	.782	.056	.000												
	N	100	100	100	100	100	100	100	100	100	100	100	100	100											
Comm Flo	Pearson correlation	-.042	.046	-.164	.429(**)	.478(**)	.388(**)	.467(**)	.474(**)	.291(**)	-.017	-.013	.575(**)	.466(**)	1										
	Sig. (2-tailed)	.675	.647	.103	.000	.000	.000	.000	.000	.003	.870	.901	.000	.000											
	N	100	100	100	100	100	100	100	100	100	100	100	100	100	100										
Rwd Sys	Pearson correlation	.025	.253(*)	-.116	.448(**)	.429(**)	.476(**)	.620(**)	.521(**)	.324(**)	.023	.159	.647(**)	.721(**)	.521(**)	1									
	Sig. (2-tailed)	.803	.011	.252	.000	.000	.000	.000	.000	.001	.819	.115	.000	.000	.000										
	N	100	100	100	100	100	100	100	100	100	100	100	100	100	100	100									
Resp	Pearson correlation	.138	-.075	-.099	.240(*)	.189	.252(*)	.474(**)	.208(*)	-.023	.097	.082	.333(**)	.254(*)	.366(**)	.277(**)	1								
	Sig. (2-tailed)	.171	.461	.326	.016	.059	.012	.000	.038	.817	.335	.417	.001	.011	.000	.005									
	N	100	100	100	100	100	100	100	100	100	100	100	100	100	100	100	100								
Conf Reso	Pearson correlation	.084	.216(*)	-.217(*)	.454(**)	.410(**)	.474(**)	.658(**)	.557(**)	.344(**)	-.020	.190	.666(**)	.681(**)	.590(**)	.632(**)	.310(**)	1							
	Sig. (2-tailed)	.407	.031	.030	.000	.000	.000	.000	.000	.000	.847	.058	.000	.000	.000	.000	.002								
	N	100	100	100	100	100	100	100	100	100	100	100	100	100	100	100	100	100							
Org Strc	Pearson correlation	-.116	.162	-.095	.027	.051	.083	.084	.050	.111	.048	-.016	.089	.115	.143	.233(*)	.035	-.036	1						
	Sig. (2-tailed)	.249	.107	.346	.788	.612	.411	.404	.621	.273	.633	.875	.379	.256	.156	.019	.730	.724							
	N	100	100	100	100	100	100	100	100	100	100	100	100	100	100	100	100	100	100						
Moti Lvl	Pearson correlation	.002	-.010	-.153	.434(**)	.486(**)	.362(**)	.428(**)	.453(**)	.280(**)	-.056	-.003	.556(**)	.418(**)	.747(**)	.502(**)	.337(**)	.477(**)	.130	1					
	Sig. (2-tailed)	.984	.924	.130	.000	.000	.000	.000	.000	.005	.582	.973	.000	.000	.000	.000	.001	.000	.196						
	N	100	100	100	100	100	100	100	100	100	100	100	100	100	100	100	100	100	100	100					
DM	Pearson correlation	-.187	.055	-.130	.396(**)	.513(**)	.362(**)	.467(**)	.429(**)	.290(**)	.033	-.136	.560(**)	.477(**)	.854(**)	.602(**)	.258(**)	.636(**)	.123	.708(**)	1				
	Sig. (2-tailed)	.063	.585	.199	.000	.000	.000	.000	.000	.003	.743	.179	.000	.000	.000	.000	.010	.000	.223	.000					
	N	100	100	100	100	100	100	100	100	100	100	100	100	100	100	100	100	100	100	100	100				
SS	Pearson correlation	.040	.161	-.223(*)	.527(**)	.565(**)	.502(**)	.577(**)	.520(**)	.332(**)	-.003	.122	.693(**)	.545(**)	.722(**)	.637(**)	.283(**)	.640(**)	.192	.545(**)	.635(**)	1			
	Sig. (2-tailed)	.696	.109	.026	.000	.000	.000	.000	.000	.001	.980	.228	.000	.000	.000	.000	.004	.000	.056	.000	.000				
	N	100	100	100	100	100	100	100	100	100	100	100	100	100	100	100	100	100	100	100	100	100			
Warmth	Pearson correlation	.154	.068	-.201(*)	.313(**)	.264(**)	.337(**)	.437(**)	.368(**)	.213(*)	-.102	.173	.445(**)	.363(**)	.569(**)	.347(**)	.555(**)	.306(**)	.146	.675(**)	.414(**)	.315(**)	1		
	Sig. (2-tailed)	.125	.503	.045	.002	.008	.001	.000	.000	.033	.313	.085	.000	.000	.000	.000	.000	.002	.148	.000	.000	.001			
	N	100	100	100	100	100	100	100	100	100	100	100	100	100	100	100	100	100	100	100	100	100	100		
Id Prob	Pearson correlation	.019	.220(*)	-.229(*)	.502(**)	.430(**)	.550(**)	.676(**)	.556(**)	.324(**)	.046	.135	.701(**)	.691(**)	.633(**)	.719(**)	.320(**)	.762(**)	.187	.536(**)	.644(**)	.775(**)	.438(**)	1	
	Sig. (2-tailed)	.852	.028	.022	.000	.000	.000	.000	.000	.001	.652	.180	.000	.000	.000	.000	.001	.000	.062	.000	.000	.000	.000		
	N	100	100	100	100	100	100	100	100	100	100	100	100	100	100	100	100	100	100	100	100	100	100	100	
OC	Pearson correlation	.015	.161	-.226(*)	.552(**)	.534(**)	.535(**)	.683(**)	.605(**)	.374(**)	.003	.324	.754(**)	.733(**)	.871(**)	.782(**)	.462(**)	.776(**)	.248(*)	.792(**)	.840(**)	.816(**)	.632(**)	.849(**)	1
	Sig. (2-tailed)	.885	.109	.024	.000	.000	.000	.000	.000	.000	.980		.000	.000	.000	.000	.000	.000	.013	.000	.000	.000	.000	.000	
	N	100	100	100	100	100	100	100	100	100	100	100	100	100	100	100	100	100	100	100	100	100	100	100	100

*Correlation is significant at the 0.05 level (2-tailed).
**Correlation is significant at the 0.01 level (2-tailed).

($P < .01$) and overall leadership effectiveness ($P < .01$). Conflict resolution motivational level, decision-making process, and support system are significantly correlated with interpersonal relation ($P < .01$), intellectual operations ($P < .01$), behavioral and emotional stability ($P < .01$), ethical and moral strength ($P < .01$), adequacy of communication ($P < .01$), operations as a citizen ($P < .01$), and with overall leadership effectiveness ($P < .01$). Warmth is significantly correlated with interpersonal relation ($P < .01$), intellectual operations ($P < .01$), behavioral and emotional stability ($P < .01$), ethical and moral strength ($P < .01$), adequacy of communication ($P < .01$), overall leadership effectiveness ($P < .011$), and with operations as a citizen ($P < .05$). Identity problem is significantly correlated with interpersonal relation ($P < .01$), intellectual operations ($P < .01$), behavioral and emotional stability ($P < .01$), ethical and moral strength ($P < .01$), adequacy of communication ($P < .01$), operations as a citizen ($P < .01$), and with leadership effectiveness ($P < .01$). Results revealed that organizational structure has no significant correlation with leadership effectiveness and its dimensions.

An overview of Table 1 reveals leadership effectiveness is also significantly correlated with organizational trust ($P < .01$). Trust in supervisor and trust in organization are significantly correlated with leadership effectiveness ($P < .01$). Interpersonal relations is moderately correlated with trust in supervisor ($P < .05$).

Intellectual operations is found to be significantly correlated with trust in organization. Ethical and moral strength is found to be correlated with trust in supervisor ($P < .05$) and trust in organization ($P < .05$).

Adequacy of communication is significantly correlated with trust in supervisor ($P < .01$). Operations as citizen is significantly correlated with trust in organization ($P < .01$).

Perceived organizational climate ($P > .05$) and its dimensions such as social support ($P > .05$), warmth ($P > .0.01$), and identity problems ($P > .05$) are found to be negatively correlated with self-efficacy. Dimensions such as organizational structure, motivational level, and decision-making were found to have no correlation with self-efficacy.

Perceived organizational climate's domains performance standards ($P < .05$), conflict resolution ($P < .05$), reward system ($P < .05$), and identity problems ($P < .05$) dimensions are found to be positively correlated with trust in organization.

Leadership effectiveness and its dimension ethical and moral strength are found to be negatively significantly correlated with self-efficacy ($P >$ .01). The other dimensions which are negatively correlated are interpersonal relations ($P > .05$), behavioral and emotional stability ($P > .05$), and adequacy of communication ($P > .05$).

Perceived organizational climate and its dimensions have insignificant correlation with job involvement.

11.2   PART II

All our findings further enabled us to find out the predictors of the leadership effectiveness, self-efficacy, interpersonal trust, and job involvement among virtual workers with regard to the perceived organizational climate and its domains.

Table 2 reveals that when the independent variable as overall perceived organizational climate and its domains entered in regression model with leadership effectiveness as criterion, overall organizational climate alone contributed 56.8% of the variance. A significant increase of .2 was obtained in R^2 when it was entered along with its domain, that is, communication flow in the regression model accounting for 59.5%. A significant increase of .604 was obtained in R^2 when it was entered along with its domain, that is, communication flow and social support in the regression model accounting for 61.6% of the variance. A significant increase of .616 was obtained in R^2 when it was entered along with domains support system and organization structure in the regression model accounting for 63.1% of the variance.

TABLE 2   Summary of Stepwise Regression Using Leadership Effectiveness As Criterion Variable

S. No.	Variables	R	R^2	Adjusted R^2	Std. error of the estimate
1.	OC	.754(a)	.568	.564	23.492
2.	OC, Comm Flo	.772(b)	.595	.587	22.851
3.	OC, Comn Flo	.785(c)	.616	.604	22.390
4.	OC, Comn Flo, SS, Org Strc	.795(d)	.631	.616	22.038

Coefficients(a)

Model		Unstandardized coefficients		Standardized coefficients	*t*	Sig.
		B	**Std. error**	**Beta**		
1.	**(Constant)**	93.147	18.227		5.110	.000
	OC	.903	.080	.754	11.350	.000
2.	**(Constant)**	95.227	17.749		5.365	.000
	OC	1.254	.157	1.047	7.967	.000
	Comn Flo	−2.311	.902	−.337	−2.563	.012
3.	**(Constant)**	89.934	17.549		5.125	.000
	OC	1.026	.184	.857	5.563	.000
	Comn Flo	−2.393	.884	−.349	−2.707	.008
	SS	2.035	.906	.246	2.245	.027
4.	**(Constant)**	108.756	19.622		5.543	.000
	OC	1.109	.186	.926	5.958	.000
	Comn Flo	−2.663	.880	−.388	−3.025	.003
	SS	2.014	.892	.243	2.258	.026
	Org Strc	−1.858	.919	−.132	−2.022	.046

[a] Dependent variable: L effect

TABLE 3   Summary of Stepwise Regression Using Self-Efficacy as Criterion Variable

Variable	*R*	R^2	Adjusted R^2	Std. error of the estimate
Id Prob	.229(a)	.053	.043	2.826

Anova(B)

Model		Sum of squares	df	Mean square	F	Sig.
1	Regression	43.443	1	43.443	5.441	.022(a)
	Residual	782.517	98	7.985		
	Total	825.960	99			

[a] Predictors: (Constant), Id Prob

[b] Dependent variable: S Effac

Table 3 depicts that when independent variable entered in the regression model with self-efficacy as a criterion, identity problem contributed 53% of the variance.

When overall perceived organizational climate was entered in the regression model with job involvement and interpersonal trust as a criterion, there was no significant relation found.

None of the domains of perceived organizational climate emerged as a significant predictor for job involvement and trust.

KEYWORDS

- organizational climate
- leadership effectiveness
- interpersonal trust
- job involvement
- decision-making process

CHAPTER 12

DISCUSSION

CONTENTS

The study was carried out to find the correlation between perceived organizational climate and other variables that is, leadership effectiveness, self-efficacy, interpersonal trust, and job involvement among virtual workers working in different organizations.

As per feasibility, results are being discussed under two broad parts:

1. Correlation between perceived organizational climate and leadership effectiveness, self-efficacy, interpersonal trust, and job involvement among virtual workers working with different organizations.
2. Find out the best predictors of leadership effectiveness, self-efficacy, interpersonal trust, and job involvement variables with regard to perceived organizational climate under study.

12.1 PART I

In order to discuss the first part of the results, the correlation matrix for the variables under study, in case of total sample ($N = 100$), Table 1 was considered.

The first hypothesis stated that there will be significant positive relationship between perceived organizational climate and leadership effectiveness among virtual workers working in different organizations.

The results reveal that in total sample of virtual workers, the perceived organizational climate is significantly and positively correlated with leadership effectiveness. It implies that the virtual workers who have positive perceived organizational climate feel effective leadership in their organization. Leadership is a key to influencing organizational behavior and achieving organizational effectiveness.

In fact from close analysis, it is found that 10 dimensions of perceived organizational climate, that is, performance standards, communication flow, reward system, responsibility, conflict resolution, motivational level, decision-making process, support system, warmth, and identity problems except organizational structure are correlated with overall leadership effectiveness and with its dimensions.

But there is a strong relationship between all other dimensions of perceived organizational climate and leadership effectiveness, which supports the first hypothesis strongly. This finding goes congruent with the line of findings and the caselet.

When artifacts are eliminated, studies of leadership succession show a strong leader's influence on the organization's performance standards. Corporate leaders play a central role in setting the ethical and moral values for their organizations. In 1976, when James Burke, head of Johnson & Johnson, challenged his management team to reaffirm the company's historic commitment to ethical behavior, he had no idea he would be asked to demonstrate that commitment in action. But when poisoned package of Tylenol appeared on store shelves, Burke did not hesitate to act on what he had pledged. The company pulled the product from the shelves at a cost of $100 million. It also offered a reward and revamped the product's package.

In the end, Tylenol recovered and is once again the leading pain medication in the United States. Perceptions of organizational climate, leadership, and group processes were aggregated within hierarchically nested work groups. Relationships across hierarchical boundaries were examined for two samples at different hierarchical levels in a military organization. Perceptions of climate were positively related across levels in both samples. There was evidence that the pattern of relationship among the other constructs was different in the two samples (Griffin & Matheiu, 2010).

The trend toward physically dispersed work groups has necessitated a fresh inquiry into the role and nature of team leadership in virtual settings. To accomplish this, 13 culturally diverse global teams from locations in Europe, Mexico, and the United States, were assembled assigning each team a project leader and task to complete. The findings suggested that effective team leaders demonstrate the capability to deal with paradox and contradiction by performing multiple leadership roles simultaneously (behavioral complexity). Specifically, it was discovered that highly effective virtual team leaders act in a mentoring role and exhibit a high degree of understanding (empathy) toward other team members. At the same time, effective leaders were also able to assert their authority without being perceived as overbearing or inflexible. Finally, effective leaders were found to be extremely effective at providing regular, detailed, and prompt communication with their peers and in articulating role relationships (responsibilities) among the virtual team members (Kayworth & Leidner, 2011).

Leadership behavior accounts for anywhere from 40–80% of the variance in many of his studies investigating factors that influence organizational climate (Ekvall, 1997).

It was investigated if the perceived effectiveness of the leader's ability to support innovation (LSI) has a significant effect on the perceived organizational climate, using a one-way analysis of variance (ANOVA) was computed. The results indicated that there was meaningful relationship between the ways that individuals perceive their organizational climate and how they observed their leaders ability to support innovation and confirms that leaders have a significant impact on the perceived organizational climate (Akkermans, Isaksen & Isaksen, 2008).

Ekwall & Ryhammar (1998) investigated a sample of 130 teachers about the creative climate in their departments, the department head's leadership, and the performance of the department in terms of creativity and productivity. The results indicated that the behavioral style of manager affects organizational results only through influencing the social climate.

The reason behind no relation between organizational structure and its dimension may be the sample, that is, virtual workers. This type of organization looks much more like a network; a network of people linked together by binding ties like deliverables and project assignments, regardless of organizational affiliations.

Organizational structure belongs to an era past; a time when vertical organizations were locked into place guided by gurus like Frederick Winslow Taylor who espoused one-to-one reporting in a top down closely knit structure with management at the top overseeing the unskilled workers at the bottom. But all of that has changed. So, organizations are suffering an identity crisis (Lojeski, 2011).

The second hypothesis stated that there will be significant positive relationship between perceived organizational climate and self-efficacy among virtual workers working in different organizations. Correlation analysis indicated that perceived organizational climate and self-efficacy among virtual workers are negatively correlated. Various possible reasons are advanced for this finding. This finding can be explained by the fact that first virtual work perceived organization climate differently than from workers working in face-to-face teams. On the basis of the findings, it can be implied that self-efficacy represents an individual's perception of their ability to plan and take action to reach a particular goal. Self-efficacy (task-specific confidence) is measured by getting efficacy ratings across a whole range of possible performance outcomes rather than from a single outcome (Cheng and Tsai, 2002). The definition and dimensionality of organizational climate has also been customized to specific contexts.

Organizational climate depends on the employee's attitude how they interpret the climate of the organization (Choudhary, 2011). It is possible that an individual is high in self-efficacy but his/her perceived organization climate is not conducive for achieving task-specific confidence. The virtual work force is comprised of people whose communications are diffused through the Internet and those who use varying frames from which to interpret and assign meaning to other's actions and words. In most situations, little in the way of context is now known in the same way it organically presented itself in firm structural formations or shared physical spaces of the past. Weaving together a whole picture about one's colleagues and mission at any given time, or determining the proper perspective from which to interpret discourse from an e-mail or other electrified communication is no longer a completely natural process.

Moreover, when organizations use old management models to try and solve new challenges and, at the same time expect individuals in social networks, tied together by electronic gadgetry, to behave exactly the same way as they did when grouped next to each other in offices where the organizational chart truly represented the working hierarchy, phantom expectations of effectiveness and worker performance emerge and usually miss the mark. The reason–the rise of virtual distance (Lojeski, 2011).

Staples, Hulland, and Higgins (1998) investigated how virtual organizations can manage remote employees effectively. The research used self-efficacy theory to build a model that predicts relationships between antecedents to employees' remote work self-efficacy assessments and their behavioral and attitudinal consequences. The model was tested using responses from 376 remotely managed employees in 18 diverse organizations. Overall, the results indicated that remote employees' self-efficacy assessments play a critical role in influencing their remote work effectiveness, perceived productivity, job satisfaction, and ability to cope. Furthermore, strong relationships were observed between employees' remote work self-efficacy judgments and several antecedents, including remote work experience and training, best practices modeling by management, computer anxiety, and IT capabilities. Because many of these antecedents can be controlled managerially, these findings suggested important ways in which a remote employee's work performance can be enhanced, through the intermediary effect of improved remote work self-efficacy.

A primary objective of organizational virtual work programs (e.g., providing the option to employees to work from home) is the reduction of employees' work–nonwork conflict and job stress. Raghuram and Wiesen-feld (2004) found preliminary evidence suggesting that virtual work is negatively related to work–nonwork conflict and job stress. They identify the work factors (clarity of appraisal criteria, interpersonal trust, and organizational connectedness) and individual factors (self-efficacy and ability to structure the workday) associated with work–nonwork conflict and find that these associations are moderated by the extent of virtual work.

Brunett, Keith, and Allan (2002) investigated the relationship between perceived organizational culture and the current general level of Internet self-efficacy as well as with two subcategories of expertise: use of course management shell software and Web authoring software. The perceived organizational dimension of "employee focus" was positively correlated with the three technology measures whereas the perceived organizational dimension of "results focus" was negatively correlated with these three measures. Although the findings are mitigated due to the small sample size, they do correspond to a general perception within academic institutions that staff are more likely to be innovative when there is support for their efforts than when there is simply pressure to produce. Overall, the findings of this research project suggested that organizations moving online need to be cognizant of the effect of different organizational climates on their staff and should take this into account when planning their change management strategies.

Luo, Huang, and Chen (2008) studied and examined the characteristics of the creative organizational climate of Chinese schools, teacher's general self-efficacy, creativity self-efficacy, and cultural self-efficacy of their own culture and analyzed the relations between these four core variables. Analyses revealed significant teaching duration difference, age difference, and a significant two-way school teaching subject interaction in creative organizational climate as well as significant gender difference in general self-efficacy.

The third hypothesis stated that there will be significant positive relationship between perceived organizational climate and organizational trust among virtual workers. The results indicated that perceived organizational climate dimension, performance standards, conflict resolution, reward system, and identity problem dimensions are positively correlated with trust in organization. The results of this study partially support this hypothesis. As virtual work climate is different from other face-to-face work environment, so the past theories and researches in virtual context do not remain unchanged and this could be one of the reason for partial correlation between the variables. The finding goes in consonance with the interview taken by the researcher:

Although, I am working in India's best MNC in IT sector but IT sector has its own complexities, everything was good but after working for so long in this reputed company, many people will be fired in month or two. X Case Senior Project Head.

No, its like Boss is keeping tab on me, checking whether I am at holiday or working. G Case, Hotel Business developer, Travel Firm.

The digital age has shifted many things. I can work from my home town also. No need to go office. My work is project based. No, Boss is not bothered whether I am at home or in market or office. This way company has no issues. Our Clients are in US so in night I can work also. B Case, IT.

Although its one of the best company in Hospitality sector, I don't think I can think more than two years. One needs to grow. Right now I am not married, so I can shift anywhere whether Bombay or Noida. A Case Hotel Solutions.

Things have changed from few years back now, at the end of day money matters, so I don't think in corporate world one marries with company whenever there is good opportunity, I can shift even if its smaller organization. H Case, ITEs firm.

Most of our organizations tend to be arranged on the assumption that people cannot be trusted or relied upon, even in tiny matters. ... It is unwise to trust people whom you do not know well, whom you do have not observed in action over time, and who are not committed to the same goals....Trust needs touch....high tech has to be balanced by high touch to build high trust organizations. Paradoxically, the more virtual an organization becomes, the more its people need to meet in person (Handy,1995).

Cultivating trust among team members in global teams has been ranked as the most difficult task by global team leaders. People trust one another more when they share similarities, communicate frequently, and operate in a common cultural context that imposes sanctions for behaving in an untrustworthy manner. Building trust is even more difficult when there is:

1. A high level of risks in the tasks.
2. A low level of interdependence between team members for accomplishing their tasks.

3. Membership that is distributed over a wide geographic area.
4. A high cultural distance between members.

Thus, it will be easier to build trust in a team with members distributed across one continent, say for instance, all the Spanish countries in Latin America, rather than a team whose members come from many different countries.

Contrary to above findings, given the growing importance and complexities of telework and the challenges associated with knowledge sharing, Golden and Raghuram (2010) investigated teleworkers and their propensity to share knowledge. They did by investigating if the relational qualities of teleworkers in the form of trust, interpersonal bond, and commitment, act to impact teleworker knowledge sharing. And also investigated how telework's altered spatial and technical interactions shape knowledge sharing, by testing the contingent role of technology support, face-to-face interactions, and electronic tool use. Results using matched data from 226 teleworkers supported the role of teleworker trust, interpersonal bond, and commitment in predicting knowledge sharing. Moreover, the impact of trust on knowledge sharing was found to be moderated by technology support, face-to-face interactions, and the use of electronic tools, whereas the impact of commitment was contingent on the use of electronic tools.

Employee's perception of organizational trust on service climate and employee satisfaction was studied by Chathoth, Mak, Jauhari, and Manaktola (2007). Multidimensional constructs of trust and service climate were developed using the literature in the trust and service management domains. Results supported that trust affects service climate and employee satisfaction, whereas service climate affects employee satisfaction in a significant way. Implications for practitioners and future research ensue, which underscored the importance of building trust and service climate to ensure employee satisfaction in hotel firms.

The results of researches (Ellonen et.al 2008, Ertu"rk 2007, Tzu–Jiun 2007, Iurato 2007, Lamsa et.al 2006, Ratnasingam 2005, Smith 2005, Politis 2003, Wang 2003, Dirks & Ferring 2001) have shown the effects of organizational trust on organizational innovation, organizational citizenship behavior, organizational commitment, motivation, organizational performance, continuance of relations, effectiveness, knowledge management and group performance, collaboration in decision-making, cooperation level and team process.

Raghuram, Garud, Weisendfeld, and Gupta (2000) explored structural factors (i.e., work independence and evaluation criteria) and relational factors (i.e., trust and organizational connectedness) as predictors of adjustment to virtual work. Additionally, age, virtual work experience, and gender as moderators of the relationships was explored. It was found that structural and relational factors are important predictors of adjustment and that the strength of the relationship is contingent on individual differences.

Alawi, Mrzooqi, and Mohammed (2007) investigated the role of certain factors in organizational culture in the success of knowledge sharing. Such factors as interpersonal trust, communication between staff, information systems, rewards, and organization structure play an important role in defining the relationships between staff and in turn, providing possibilities to break obstacles to knowledge sharing. The research findings indicated that trust, communication, information systems, rewards, and organization structure were positively related to knowledge sharing in organizations.

Sullivan, Peterson, Kameda, and Shimada (1981) investigated whether the manner in which conflicts are resolved in Japanese-American joint ventures in Japan influences the level of future mutual trust. Japanese managers perceived a higher level of future trust when disputes are resolved through conferral, except when an American is in charge of operations. Then they designated contracts requiring binding arbitration.

Trust development in virtual teams may be more difficult in the absence of face-to-face contact (McDonough, Kahn, & Barczak, 2001). Studies have identified the difficulty of communicating in virtual teams because of the lack of media richness (Watson-Manheim & Belanger, 2002). Computer-mediated communication depersonalizes the interaction, so there is a greater focus on the actual words in the message (Sproull & Kiesler, 1991). If the communication only serves to report or inform, the possibility of misunderstanding may be low. But when communications go beyond simple reporting to task allocation and negotiation, the receiver may misinterpret the meaning of the message (Furomo and Pearson, 2007).

Wilson (1993) believes that although trust is a significant concept for the study, it is a topic which has different interpretations. Lewicki et.al (1998: 443) defines trust as "The perception of one about others, decision to act based on speech, behavior and their decision." Sashkin (1990:6) defines trust as "the confidence which employees feel about management and their belief toward what management tell them." Mayer et.al (1995) believes that trust is "the tendency of a group to susceptibility toward the actions of other group, it is expected that group will do a special action which is important in the view of the one who trusts regardless of supervision and control of the group." Mishra (1996) defines organizational trust as a unidirectional tendency toward susceptibility to other party based on this expectation or believe that the other party is reliable, open, and trustable.

The fourth hypothesis of this study stated that there will be significant relationship between perceived climate and job involvement among virtual workers. Results indicated that perceived organizational climate and job involvement are not significantly correlated. Various possible reasons can be advanced for this finding. The results are in line with researcher's observations taken during interview schedule with virtual workers:

I have both opinions. Its better to be in office I think. Atleast one has some schedule at 9 am you have to get ready and in evening atleast by 8-9 you will be back at home. But right now I don't have any schedule I have to work anytime anywhere its like 24x 7. My Efficiency decreases in home. C Case IT Sector.

We can save time and energy by Working virtually but it depends on circumstances, I have small kids and my wife is also working in bank so at times its convenient for me to take care of kids as I work from home but it has its distractions at times but yes overall I am happy as I am a family man now. At this stage I am not very aggressive in work earlier yes I was. H Case, GM Hospitality.

I personally don't prefer to be virtual worker. As there are so many distractions at home and in any case of any escalated mail, I cannot ask or control my team from home. My team members can work from home, I have no objections but I don't. Job Involvement is not much in comparison to my first company. D Case IT Sector.

Integrated model theory covers the dispositional approach in which job involvement is viewed as dependent on individual personality. The influence exerted by some stable personal characteristics such as age, gender, marital status, external and internal control features, job seniority, dwelling locations, the intensity of high-level work demands in terms of time and responsibility, and the protestant work ethics will ensure individuals hold different work attitudes and behavior(Sekaran & Mowday, 1981). In a situational determined approach, job involvement can be viewed as the personal attitude toward the particular job.

Technology usage is also one factor which affects job involvement. Jain and Rathore (2011) investigated stress due to technology on software professionals and found that continuous stress affects the person's job involvement negatively. And virtual workers work through computers or telephones only.

This can be explained by the following studies where job involvement gets affected by other factors also.

When virtual distance is relatively high, the following critical success factors significantly degrade:

- Job satisfaction drops off by over 80%.
- Innovation falls by over 90% and competitive advantage is severely impacted.
- On-time/on-budget project performance suffers by over 50%.
- Goal and role clarity decline by over 60% (Lojeski, 2011).

Utilizing panel data for three age groups from the 1972-73 and 1977 Quality of Employment Surveys, Lorence and Mortimer (1985) investigated job involvement through the life course and found that job involvement is quite volatile in the initial stage of the work career, it becomes more stable, supporting the aging stability hypothesis as workers grow older.

Dienhart and Gregoire (1993) examined the relationship of customer focus and job satisfaction, job involvement, and job security for quick service restaurant employees. Results indicated that job satisfaction, job involvement, and job security do tend to predict customer focus for restaurant employees, and consequently, increasing job satisfaction, job involvement, and job security may improve an employee's customer focus. Having a better understanding of employees and factors which affect their focus on the services they provide to others is critical for restaurant managers as they hire and train employees.

The fifth hypothesis of this study stated that there will be correlation between different dependent variables, that is, leadership effectiveness, self-efficacy, interpersonal trust, and job involvement.

Out of all, leadership effectiveness is correlated with trust and this can be explained by leadership concept itself and trust is one of the emerging issue in leadership today. It is clear that without trust we cannot build leadership effectiveness as follower's role models and inspirations are leaders only. Trust is the willingness to be vulnerable to the actions of another. This means that followers believe that their leader will act with the follower's welfare in. Moreover, the digital world has led many of us to lead compartmentalized lives and a new leadership is the requirement for today's virtual organizations which is authentic leadership which requires leaders to develop authentic relationships with followers. These three related types of behaviors are:

1. Transparency, openness, and trust
2. Guidance toward worthy objectives
3. Emphasis on follower development

Authentic leadership proponents suggest that authentic leaders form trusting relationships which is very essential for this digital age organizations.

An analysis of the performance of 317 bank employees confirmed the positive relationship between a high employee-supervisor leader–member exchange (LMX) and high job performance. This study also showed that the relationship was maintained even when the employees and the supervisor were geographically remote from one another; that is, when the employees worked in a different office or a different city from their direct supervisor. The researchers concluded that the mutual trust and respect of a high quality LMX relationship overcame physical distance so that employees and supervisors worked jointly to accomplish organizational goals (Howell & Hall-Merenda, 1999).

Leadership and trust relationship is quite old. Questionnaire data was obtained from 102 hospital employees and 26 supervisors. The more the LMX were based on positive interpersonal interactions such as mutual trust, loyalty, and respect, the more the subordinates were likely to engage in behaviors to help coworkers and the organization (behavior in addition to their duties) (Setton, Bennett & Liden, 1996).

Wech (2002) and Steward (2004) investigated the relationship between leader–employee interaction and organizational trust and concluded that an organizational trust environment should be created for achieving a high-quality leader–employee interaction. In their studies, both Lester and Brower (2003) and Joseph and Winston (2005) obtained findings which suggest that an environment of trust should be created in the organizations for attaining efficient leadership. Milligan (2003) studied the effects of organizational trust on leadership behaviors and commitment to organization.

Casimir, Waldman, Bartram, and Yang (2006), provided a cross-cultural comparison of the mediating effects of trust in the leader on the relationship between the in-role performance of followers (as rated by their leaders) and two types of leadership: transactional and transformational. Participants were 119 full-time Australian followers and 122 full-time Chinese followers. Australian followers reported higher levels of trust in their leaders than did Chinese followers. Culture moderated the mediation effects of trust on the *leadership–performance* relationship. The findings highlighted the need to consider the cultural context within which leadership occurs when attempting to understand mediated relationships with performance outcomes.

Servant leadership, trust, team commitment, and demographic variables were measured among 417 sales persons in the automobile industry. Pearson product moment correlation, multiple regression, *t*-tests and ANOVA were used to analyze the data. Strong relationships were found among servant leadership, trust, and team commitment (Danhauser & Boshoff, 2006).

Joseph and Winston (2005) explored the relationship between employee perceptions of servant leadership and leader trust, as well as organizational trust. Perceptions of servant leadership correlated positively with both leader trust and organizational trust. The study also found that organizations perceived as servant-led exhibited higher levels of both leader trust and organizational trust than organizations perceived as non-servant led.

Gomez & Rosen (2001) asserted that trust in group leadership is positively associated with employee perceptions of the group and a better communication with managers. This leads to higher quality relationships, competence, and feelings of empowerment. This is supported by Fox, Rejeski, and Gauvin (2000) who showed that leader trust in conjunction with a supportive group environment lead to greater individual enjoyment of the group. In addition, findings from Hare and O'-Neil (2000) indicated that frustration and lower morale often resulted from mistrust between the leader and the group as well as from undefined leader/follower roles.

Hassan and Ahmed (2011) examined how authentic leadership contribute to subordinates' trust in leadership and how this trust, in turn, predicts subordinates' work engagement. A sample of 395 employees was randomly selected from several local banks operating in Malaysia. Standardized tools such as ALQ, OTI, and EEQ were employed. Results indicated that authentic leadership promoted subordinates' trust in leader, and contributed to work engagement. Also, interpersonal trust predicted employees' work engagement as well as mediated the relationship between this style of leadership and employees' work engagement.

Mathebula (2004) studied the relationship between organizational commitment, leadership style, human resource management practices and organizational trust in 246 employees from 11 universities in South Africa and found significant correlations between trust and human resource management, trust and organizational commitment, leadership style and trust. However, multiple regression analysis indicated weak predictions of organizational commitment by different independent variables.

Ponnu and Tennakoon (2009) explored the impact of ethical leadership behavior on employee attitudinal outcomes such as employees' organizational commitment and trust in leaders. The study uses primary data collected from 172 intermediate managerial level employees from the corporate sector in Malaysia. Results indicate that ethical leadership behavior has a positive impact on employee organizational commitment and employee trust in leaders.

Virtual teams enabled by information and communications technologies (ICT) are increasingly being adopted not only by for-profit organizations but also by education institutions as well. Wu, Yang, and Tsou (2008) investigated what contributes to the success of virtual learning teams. Specifically, they examined the issue of leadership in virtual learning teams. The study first reviewed the current literature on teams, leadership, and trust then proposed a framework of team effectiveness of virtual learning teams. A field study was conducted to investigate the influence of several independent variables including diversified leadership roles, leadership effectiveness, team trust, and propensity to trust. It was found that diversified leadership roles influences both leadership effectiveness and team trust; both leadership effectiveness and propensity to trust influence team trust, and team trust in turn directly impacts team effectiveness. In addition, team trust mediated the relationship between leadership effectiveness and team effectiveness.

Dirks and Ferrin (2002) examined the findings and implications of the research on trust in leadership that has been conducted during the past four decades. First, the study provided estimates of the primary relationships between trust in leadership and key outcomes, antecedents, and correlates ($k = 106$). Second, the study explored how specifying the construct with alternative leadership referents (direct leaders vs. organizational leadership) and definitions (types of trust) results in systematically different relationships between trust in leadership and outcomes and antecedents. Direct leaders (e.g., supervisors) appeared to be a particularly important referent of trust.

In continuation of fifth hypothesis, it was found that leadership and self-efficacy are significantly negatively correlated. The reason may be virtual workers as a sample of study, this can be supported by other studies done on virtual workers where conflicting results on other variables also. Because virtual work is altogether different work scenario from traditional face-to-face work offices, behavior in organization changes due to interactions among individuals, groups, and organization itself, but when whole organization structure gets changed and in fact there is no affiliation or belonging as such, how past researches done on face-to-face traditional teams can be implemented. According to Golden, Veiga, and Simsek (2008), the literature on telecommuting and work–life conflict is equivocal. In some cases, it was suggested that the alternative work schedules reduces work–life conflict; in other cases the indication is that it increases such conflict. Using a sample of 454 professional, the researchers conducted a survey to more closely examine the dynamics of telecommuting and work–family conflicts.

In this study, the researchers expected that more telecommuting would be associated with lower reported work–family conflict, or with less interference of work on family matters. This is because of the flexibility of telecommuting in allowing people to deal emotionally and directly with family responsibilities. Data confirmed this hypothesis. A second hypothesis was that more telecommuting would be associated with greater reported family–work conflict, or with more interference of family on work matters. This is the result of the increased strains of knowing that telecommuting allowed people to give more time and emotional energy to family matters. Data also confirmed this hypothesis.

Also included in the study was an attempt to understand how moderator variables, such as household size, influence the prior relationships. One moderating hypothesis was that any negative relationship between telecommuting and work–family conflict would decrease more slowly with increasing household size. This was not supported. Although data showed the expected direction, they were not statistically significant. Another moderating hypothesis was that any positive relationship with family–work conflict would increase at a faster rate with increasing household size. The researchers concluded that more needs to be learned about tele-commuting trade-offs as participants attempt to balance work and family responsibilities. They call for more research and ask for caution in terms viewing telecommuting as a panacea for work–life issues in workplace (Schermerhorn, Jr, Hunt & Osborn, 2010)

12.2   PART II

In order to examine the extent to which weighted combination of various variables included in the study predicts the criterion variable, stepwise multiple regression was applied. Stepwise multiple regression was employed to identify the factors that account for maximum proportion of the variance in overall perceived organizational climate and its dimensions, the criterion variable, and to eliminate those that do not make additional contribution to the variables already in question. In the present study, there were four predictor variables, that is, leadership effectiveness, self-efficacy, interpersonal trust, and job involvement.

Multiple regression analysis (Table 3) for total sample of virtual workers suggests that only perceived organizational climate and its three dimension , communication flow, social support, and organizational structure were the predictors among all the criterion.

Overall perceived organizational climate emerged as best predictor of leadership effectiveness. This finding may be explained in terms of the interaction among the components of perceived organization climate and criterion variable.

Renato Tagiuri (1968) defines organizational climate as a relatively ending quality of the internal environment that is experienced by the members, which influences their behavior and can describe in terms of values of a particular set of characteristics of the organization. Kurt Lewin argued that different leadership styles affect organizational climate (Choudhary, 2011).

According to Litwin and Stringer, leadership style is a critical factor for the quality of any organizational climate. In fact, a manager's behavior accounts for about 70%of the variability of climate. Good managers lower anxiety, creating a sense of confidence and security. By communicating realistically, they build trust. And they acknowledge urgency, focusing teams on the most important projects and goals. All these actions engage employees in their work and connect them to the larger organization (Atkinson & Frechetter, 2009).

The relationship between organizational climate and leadership styles of the managers of physical education organization in Isfahan province was studied. The results suggested there was significant relationship between organizational climate and leadership behavior of the managers. The correlation coefficient indicated a significant positive relationship between the autocratic leadership and closed organizational climate and between democratic leadership style and open organizational climate at the 0.01 significance level. Further, a significant positive correlation was observed between the dimensions of organizational climate (role, reward, and communication) and leadership styles of managers (Eshraghi, Harati, Ebrahimi & Nasiri, 2011).

Lewin and his associates characterized leadership within the clubs as corresponding to one of the three styles (autocratic, democratic, or laissez faire). These styles determined the .social climate within the clubs, which led in turn to particular behavior repertoires displayed by the boys. Benjamin Schneider (1975) defined organizational climate as a mutually agreed internal (or molar) environmental description of an organization's practices and procedures.

Communication flow dimension of perceived organizational climate is another dimension which emerged as predictor for leadership effectiveness which means if more and good communication flow is in context of perceived organization climate, the more effective leadership. This result is in line with various researches and theories of leadership. *Effective leadership is still largely a matter of communication. Alan Axelrod. Elizabeth I, CEO.*

Communication competence is a prerequisite for effective leadership. One hundred and fifty-one employees of nine organizations rated their immediate supervisor's communication competence and using a three dimensional integrated model of leadership categorized that supervisor's perceived leadership effectiveness. Each of the leadership dimensions, as well as the model as a whole, was highly correlated with competent communication by the supervisor (Flauto, 1990).

Vries, Pieper, and Ostenveld (2009) investigated the relations between leaders' communication styles and charismatic leadership, human-oriented leadership (leader's consideration), task-oriented leadership (leader's initiating structure), and leadership outcomes. It was found that charismatic-and human-oriented leadership are mainly communicative, while task-oriented leadership is significantly less communicative. The communication styles were strongly and differentially related to knowledge sharing behaviors, perceived leader performance, satisfaction with the leader, and subordinate's team commitment. Multiple regression analyses showed that the leadership styles mediated the relations between the communication styles and leadership outcomes. However, leader's preciseness explained variance in perceived leader performance and satisfaction with the leader above and beyond the leadership style variables.

Communication is perhaps one of the greatest challenges facing managers and leaders today. Clearly articulating ideas and expectations to employees is vital to the productivity and the longevity of an organization. Furthermore, the style in which the communication is delivered has an influence on the satisfaction levels of employees. Research has discovered that there are many different styles in which a leader may communicate with employees. Research has provided several methods that aid in determining which style is the most appropriate for any given circumstance. Research has demonstrated how appropriate and effective communication is used to promote organizational health. Furthermore, research has demonstrated how inappropriate communication may decrease employee satisfaction. Finally, research has provided methods to aid in improving communication styles and delivery (Hicks, 2011).

Through effective communication, leaders lead. Good communication skills enable, foster, and create the understanding and trust necessary to encourage others to follow a leader. Without effective communication, a manager accomplishes little. Without effective communication, a manager is not an effective leader.

A leader must be able to communicate effectively. When CEOs and other senior executives in all industries and countries are asked to list the most important skills a manager must possess, the answer consistently includes–good communication skills. Managers spend most of their day engaged in communication; in fact, older studies of how much time managers spend on various activities show that communication occupies 70–90% of their time every day (Mintzberg, 1973; Eccles & Nohria, 1991). With cell phones, e-mail, text messaging, if that same study were done today, it would yield even higher percentages. The sheer amount of time managers spend communicating underscores how important strong communication skills can be for the manager desiring to advance to leadership positions; thus, mastering leadership communication should be a priority for managers wanting their organizations or the broader business community to consider them leaders. In fact, being able to communicate effectively is what allows a manager to move into a leadership position. An early Harvard Business School study on what it takes to achieve success and be promoted in an organization says that the individual who gets ahead in business is the person who "is able to communicate, to make sound decisions, and to get things done with and through people" (Bowman, Jones, Peterson, Gronouski, & Mahoney, 1964). By communicating more effectively, managers improve their ability to get things done with and through people. Also, managers can improve their ability to project a positive ethos by building a positive reputation, improving their professional appearance, projecting greater confidence, and learning to communicate more effectively (Barrett, 2006).

Generally, a savvy leader's success is directly tied to his or her ability to focus on the business fundamentals–the daily blocking and tackling that every company must master to be a winner in its field. Strong, effective leaders stress fundamentals like discipline, accountability, strategic alignment, managing to his or her values, and empowering employees. Additionally, these leaders have mastered the six basic functions of management: leading, planning, organizing, staffing, controlling, and communicating. But what is the one golden thread tying all those functions together— and the most important key to great leadership? Clear communication (Froschheisner, 2012).

Leadership, change management, and communication are so intimately linked that it is really not possible to be successful at the first two without well-developed communication skills. Part of success in managing organizational change lies in the manager's ability to choose the right channels of communication that match the context and to phrase the messages properly (Bacal, 2012).

Lee and Lin (1999) explored the relationships among superior's leadership style, employees' communication satisfaction, and leadership effectiveness. Through both Pearson correlation analysis and path analysis of the data, the researchers managed to discuss the relationships among superior's leadership style, employees' communication satisfaction and leadership effectiveness. Meanwhile, analysis of variance was employed to determine the differences between superior's leadership style, employees' communication satisfaction and leadership effectiveness in terms of demographic variables. There was significant correlation between leadership style and employees' communication satisfaction, between employees' communication satisfaction and leadership effectiveness, and between leadership style and leadership effectiveness.

Besides communication flow dimension of perceived organizational climate, support system is another dimension which emerged as predictor for leadership effectiveness which means the more support from organization in context of perceived organization climate, the more effective leadership. This finding is in alignment with the base of various theories on leadership.

According to cognitive resource theory, plans and decisions cannot be implemented unless the group compiles with the leader's directives. Therefore, the correlation between a leader's cognitive resources and group performance will be higher when the group supports the leader than when it does not.

A disjointed group is less likely to be effective than a cohesive group. So we can say that getting leadership effectiveness group cohesion is necessary.

The LMX model deals with the ways in which the leader–follower relationship affects the leadership process.

Overall, research has shown that groups who are confident in their leadership show increased confidence in their abilities as a group, which predicts effectiveness (Shultz, Shultz, 2004).

It has been argued that transformational leaders increase group effectiveness by empowering followers to perform their job independently from the leader, highlight the importance of cooperation in performing collective tasks, and realign followers' values to create a more cohesive group. A study was conducted to examine whether transformational leadership would be positively related to followers' perceptions of empowerment, group cohesiveness, and effectiveness. Forty-seven groups from four Korean firms participated in this study. Results of partial least squares analysis indicated that transformational leadership was positively related to empowerment, group cohesiveness, and group effectiveness. Empowerment was positively related to collective efficacy, which in turn was positively related to group members' perceived group effectiveness (Jung and Sosik, 2002).

A longitudinal laboratory experiment was conducted by Sosik, Avolio, and Khai (1997) to evaluate the effects of leadership style (transactional vs. transformational) and anonymity level (identified vs. anonymous) on group potency and effectiveness of 36 undergraduate student work groups performing a creativity task using a Group Decision Support System (GDSS). GDSS are interactive networks of computers for generating solutions to unstructured problems. Results indicated that GDSS anonymity amplified the positive effect of transformational leadership on group potency relative to transactional leadership in the group writing session of the task. GDSS anonymity also increased the effect of transformational leadership relative to transactional leadership on group effectiveness.

Organizational structure domain of perceived organizational climate emerged as predictor for leadership effectiveness which means that organization structure impacts and increases leadership effectiveness. This finding goes in consonance with studies and theories of leadership. Authoritarian leaders cannot give results in flat or horizontal organizational structure. As most virtual organizations have no as such organizational structure, it certainly affects leadership effectiveness in these organizations and that is where difference between managers and leaders are seen. Mostly virtual workers work as self-managed teams, in this leaders become effective when they behave as non-traditional leaders which are called "coordinators" or "facilitators." This is the reason leadership theorists believe there is no single leadership style which is effective for every kind of organizational climate. Leaders should be chosen who challenge the organizational culture when necessary, without destroying it (Nelson, Quick, Khandelwal, 2011).

Building upon the culturally endorsed implicit theory of leadership, Huang, Rode, and Schroder (2010) investigated the moderating effects of national culture on the relationship between organizational structure and continuous improvement and learning. They proposed that the relationship between organic organizations (characterized by flat, decentralized structures with a wide use of multifunctional employees) and continuous improvement and learning will be stronger when national cultural endorsement for participative leadership is high. They further proposed that organizational group culture will moderate the relationship between organizational structure and continuous improvement and learning, but that these moderation effects will be stronger in national cultures with low endorsement of participative leadership. Empirical analysis of secondary survey data collected from 266 manufacturing plants operating in three industries and located in nine countries representing a diverse set of geographical regions provided support for the hypotheses. Overall findings indicated that, to fully realize the relationship between organic structures and continuous improvement and learning, managers must actively assess the extent to which the national culture endorses participative leadership. In cases where this endorsement is weak, managers should consider the extent to which the organizational culture will provide alternative support for the relationship.

Kahn, Wolfe, Quinn, and Snoek (1964) and Shivers-Blackwell (2004) asserted that managers' perceptions of organizational context and personality influence their interpretations of leadership role requirements. The follow-up study posited that managers' perceptions of organizational structure and culture influence how they interpret their leadership role requirements. Furthermore, locus of control and self-monitoring are proposed to moderate this relationship. One hundred and eighty-six managers were surveyed. Results indicated that there is a relationship between managers' interpretations of organizational context and their perceived transactional and transformational role requirements(Shivers-Blackwell, 2006).

Organizational structure institutionalizes how people interact with each other, how communication flows, and how power relationships are defined (Hall, 1987). The structure of an organization reflects the value-based choices made by the company (Quinn, 1988; Zammuto and O'Connor, 1992); it refers to how job tasks are formally divided, grouped, and coordinated. Quinn's (1988) competing values model shows how different value orientations of organizations can influence structure. One dimension of value systems that is related to structure is the control-flexibility dimension (Quinn, 1988; Zammuto and Krakower, 1991). Control-oriented value systems try to consolidate management control by centralizing decision-making in managerial hands and decreasing employee discretion and flexibility. And leadership is all about influencing subordinates, so in centralization leader will be authoritative person only whereas in decentralization, leaders emerge and we will find transformational leadership more effective.

A 20-item scale for values-based leadership was developed and looked at its relationship with transformational leadership and two dimensions of organizational structure–formalization and decentralization–using a sample of 100 employees of a leading software consulting firm in India.

Results showed that transformational leadership and values-based leadership are positively related to each other and that both are positively related to decentralization. The hypothesis that formalization would be negatively related to both the leadership variables was not supported. Results also revealed that when values-based leadership is controlled for, transformational leadership is no longer related to decentralization (Garg & Krishnan, 2003).

Identity problem as a domain of perceived organizational climate emerged as the best predictor for self-efficacy. It shows that if identity problem increases, self-efficacy also decreases. The reason may be virtual work setting. There is no belongingness or organization citizenship in this kind of work environment as there is no office, no colleagues, and no leaders. In fact, some believe due to globalization and technology advancements in coming years, leaderless business organization can be a workable reality. But so far our bouillabaisse of change includes growing organizational identity problems and crisis by increasingly high level of high virtual distance. And yet, leaders at Google, the world leader in providing virtual content and information, believe that people work best when seated together side by side.

Moreover, everything gets changed except human behavior in this digital world. So, when there is identity problem in organization certainly as in virtual settings, self-efficacy will get effected because if we analyze general self-efficacy is a broader term ("I believe I can perform well in just about any part of the job") than task-specific self-efficacy which describes a person's belief that he or she can perform specific task. Self-efficacy is hypothesized to be an important determinant of action, given the appropriate level of skill and performance (Locke & Latham, 1990).

Self-efficacy is defined as one's belief in their ability to perform a given behavior (Bandura, 1977; Wood & Bandura, 1989).

Whether task and self-regulatory self-efficacy (scheduling and barriers), and self-identity predict maintenance physical activity was examined by Strachan, Woodgate, Brawley, and Tse (2005). Sixty-seven maintenance runners completed self-efficacy and self-identity measures and, 4 weeks later, recalled their physical activity. Two multiple regression analyses indicated that when combined with self-identity in independent models, both forms of self-regulatory self-efficacy predicted running frequency. A model consisting of task self-efficacy and self-identity significantly predicted running duration. In an extreme self-identity group MANOVA, the high group showed more favorable social cognitions and behavior than the low group.

Jakubowski and Dembo (2004) examined the relationship between academic self-regulation, self-efficacy, and two student self-belief systems, identity style and stage of change, for 210 college students enrolled at a private research university. High scores on the informational identity, contemplation stage, and action stage subscales and low scores on the diffuse/avoidant identity subscale were correlated with high self-regulation scores. The degree that the students have invested effort in establishing their identity as students (informational identity style) and their willingness to improve their study skills (action stage of change) significantly increased the proportion of variance explained in students' self-regulation scores.

With a sample size of 128 deaf or hard-of-hearing participants, Durand (2008) investigated the relationships between variables related to cultural identity, perceived career barriers, and self-efficacy for career decision-making. Career self-efficacy and perceived career barriers were examined using the Career Decision Self-Efficacy Scale and Career Barriers Inventory Revised, respectively. Cultural identity was examined using the revised Deaf Identity Development Scale. Data gathered were analyzed using correlational analyses, one-way ANOVAs, and simultaneous multiple regressions. The study expanded on previous research related to deaf/hard-of-hearing individuals' identities and career development by drawing from a sample of primarily employed adults with a wide age span (20–87 years), who were predominately bicultural identified. Statistical significance was not found in the relationship between career-related self-efficacy and perceived career barriers. Regarding preference for mode of communicating, no significant differences were found in relation to either of the examined career-related variables. Age of participants was significantly correlated with three of the four deaf identity stages. Deaf identity was a significant predictor of variance for career decision-making self-efficacy.

Creativity at work by considering a new construct, creative personal identity, in conjunction with creative self-efficacy and a problem-solving strategy was examined. Results of a field study suggested that creative personal identity explained variance in creativity at work above and beyond creative self-efficacy, but that the two did not interact. Results also indicated support for the interaction of the self-concept and a problem-solving strategy. The positive relationship between creative personal identity and creativity at work was stronger when individuals applied nonwork experiences in efforts to solve work-related problems (Jaussi, Randel & Dionne, 2007).

The relationship between the social cognitive construct of career decision-making self-efficacy and the outcome variables of vocational identity and career exploration behaviors in a sample of 72 urban African American high school students was explored by Gushue, Scanlan, Pantzer, and Clarke (2006). The results indicated that higher levels of career decision-making self-efficacy were related to both a more differentiated vocational self-concept and to greater engagement with career exploration activities.

The factor structure of a recently developed Norwegian scale for measuring teacher self-efficacy and partly to explore relations between teachers' perception of the school context, teacher self-efficacy, collective teacher efficacy, ,and teachers' beliefs that factors external to teaching puts limitations to what they can accomplish, was tested. Participants were 2249 Norwegian teachers in elementary school and middle school. The data were analyzed by means of structural equation modeling using the AMOS 7 program. Teacher self-efficacy, collective efficacy, and two dimensions of burnout were differently related both to school context variables and to teacher job satisfaction (Skaalvik & Skaalvik, 2009).

KEYWORDS

- ANOVA
- leadership effectiveness
- self-efficacy
- interpersonal trust
- job involvement

CHAPTER 13

SUMMARY AND CONCLUSION

CONTENTS

13.1   OBJECTIVES

- To study whether perceived organizational climate is related to leadership effectiveness among the virtual workers of different organizations.

1. To examine the relationship of perceived organizational climate with self-efficacy among virtual workers of different organizations.
2. To find out whether perceived organizational climate is related to trust among virtual workers of different organizations.
3. To find out whether perceived organizational climate is related to job involvement among virtual workers in different organizations.
4. To find out the intercorrelation among the different criterion variables, that is, leadership effectiveness, self-efficacy, trust, and job involvement.
5. To find out the best predictors of leadership effectiveness, self-efficacy, trust, and job involvement with regard to perceived organizational climate.

13.2   HYPOTHESES

The study intends to test the following hypotheses:

1. There will be significant positive relationship between perceived organizational climate and leadership effectiveness among virtual workers of different organizations.
2. There will be significant positive relationship between perceived organizational climate and self-efficacy among virtual workers of different organizations.
3. There will be significant positive relationship between perceived organizational climate and trust among virtual workers of different organizations.
4. There will be significant positive relationship between perceived organizational climate and job involvement among virtual workers of different organizations.
5. There will be positive relationship between the different criterion variables under study.
6. Considerable part of variance in independent variables (predictors), that is, perceived organizational climate will contribute to the

dependent variables (criterions) that is, leadership effectiveness, self-efficacy, trust, and job involvement among virtual workers of different organizations.

13.3 VARIABLES

Predictor variables: Perceived organizational climate and its domains: Performance standards, communication flow, reward system, responsibility, conflict resolution, organizational structure, motivational level, decision-making process, support system, warmth, and identity problems.

Criterion variables:
   i.   Leadership effectiveness and its domains, interpersonal relations, intellectual operations, behavioral and emotional stability, ethical and moral strength, adequacy of communication, operations as a citizen
   ii. Self-efficacy
   iii. Trust and its domains, trust in supervisor and trust in organization
   iv. Job involvement

13.4 RESEARCH DESIGN

Fisher (1951) remarked, "If the design of an experiment is faulty, any method of interpretation, which makes it out to be decisive, must be faulty too." Pointing out the importance, Kerlinger (1973) remarked that the chances of arriving at accurate and valid conclusions are better with social designs than with unsocial ones. If the design is faulty, one can come to no clear conclusions. For the present study, correlation design and multiple regression analysis was used.

13.4.1 CORRELATIONAL DESIGN:

The important defining features of correlational research are that the researchers do not directly manipulate the variable under study. As a general statement, we may say that research is likely to be correlational in nature, whenever all the variable under study concern properties of the subject which are either inherent to the subject (e.g., age, sex, intelligence, etc.) or the result wholly or in part of prolonged experienced (e.g., aspects of personality). The coefficient of correlation tells us the way in which two variables are related to each other. How the change in one is influenced by the change in other may be explained in terms of direction and magnitude of these measures.

The present investigation deals with the interrelationship between perceived organizational climate, leadership effectiveness, interpersonal trust, self-efficacy, and job involvement among virtual workers in different organizations. So the investigator thought it proper to adopt correlation design.

CORRELATIONAL DESIGN
Order of administration of measurement devices: Randomly

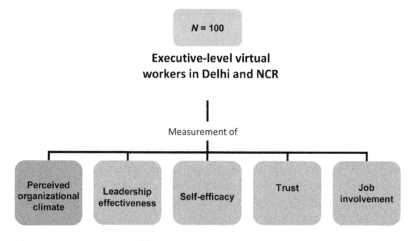

13.5   MEASURES

The following tools and techniques were used to measure different variables:

1. Organizational Climate Inventory–Chattopadhaya & Aggarwal (1976)
2. Leadership Effectiveness Scale–Taj Hassan (2001)
3. Generalized Perceived Self-efficacy Scale–Schwarzer & Borm (1997)

4. Organizational Trust Inventory–Nyhan & Marlowe (1997)
5. Measurement of job and work involvement–Kanungo (1982)

13.6 STATISTICAL ANALYSIS

Correlation, stepwise multiple regression analysis was applied to the data to draw inferences.

13.7 MAJOR FINDINGS

I. Correlation analysis between perceived organizational climate, leadership effectiveness, self-efficacy, interpersonal trust, and job involvement among virtual workers in different organizations:

1. Correlation between perceived organization climate and leadership effectiveness has been found significant among virtual workers
2. Correlation between perceived organization climate and its dimensions like social support, warmth and identity problems is found negatively correlated with self-efficacy among virtual workers.
3. Correlation between overall perceived organization climate and trust has not been found significant among virtual workers. But perceived organizational climate's domains performance standards, conflict resolution, reward system, and identity problems dimensions are found positively correlated with trust in organization.
4. Correlation between perceived organization climate and job involvement has not been found significant among virtual workers.
5. Correlation between overall leadership effectiveness and overall trust has been found to be positive significantly among virtual workers. Trust in supervisor and trust in organization with leadership effectiveness has been found significantly correlated. Interpersonal relations are moderately correlated with trust in supervisor (dimension of trust). Intellectual operations dimension of leadership effectiveness has been found to be significantly correlated with trust in organization (dimension of trust). Ethical and moral strength (dimension of leadership effectiveness) is found to be correlated with trust in supervisor and trust in organization (dimensions of trust). Adequacy of communication

is significantly correlated with trust in supervisor. Operations as citizen are significantly correlated with trust in organization.

6. Leadership effectiveness and its dimension ethical and moral strength with self-efficacy are found negatively significantly correlated among virtual workers. The other dimensions which are negatively correlated are interpersonal relations, behavioral and emotional stability, and adequacy of communication.

II. Multiple regression analysis between perceived organizational climate and other variables under study:

1. Multiple regression confirms that the overall perceived organizational climate was considered as the best predictor for enhancing leadership effectiveness.

2. Multiple regression confirms that the predictive value of communication flow, support system, and organizational structure domains of perceived organizational climate enhance prediction of leadership effectiveness.

3. Value of multiple regression confirms that the identity problem domain of perceived organizational climate was found as an important predictor for determining self-efficacy of the virtual workers.

KEYWORDS

- **communication flow**
- **reward system**
- **conflict resolution**
- **interpersonal relations**
- **self-efficacy**

CHAPTER 14

LIMITATIONS, SUGGESTIONS, AND IMPLICATIONS

CONTENTS

14.1   LIMITATIONS AND SUGGESTIONS

The study was designed scientifically and was conducted using suitable techniques. Accuracy and perfection cannot be possibly drawn in a single research work, as all the factors are difficult to be taken into stride by an individual in a stipulated amount of time. The present research too had limitations. Moreover, the research topic is very intriguing because perceived organizational climate is so interwoven of various complex factors that it becomes difficult to control one and study another. The researcher felt that if certain limitations had been considered, it would have yielded more perspective picture of perceived organizational climate with other variables among virtual workers. Some of the limitations are being stated here in brief:

1. In the study, if the purposive sample of virtual workers had been increased, it would have certainly enlightened more on variables studied.
2. In this study, the respondents were taken from Delhi and NCR only. Wide geographical area covered would have shown light in the proper relation of various variables.
3. Comparison with face-to-face teams could have been done for better findings.
4. Cross-cultural research would have definitely contributed to the research as the results could be then generalized on a wider population.
5. The study would have yielded more accurate results if perceived organizational climate was dealt with the personal and cultural background of virtual workers.
6. In the present study, demographic variables such as socioeconomic status, gender, etc. may also be important.
7. Some of the questionnaires used in the study were very lengthy as the subjects complained of the monotonous effect that led to insignificant results in some important variables.
8. Different industries or sectors of virtual workers would have yielded more clarity in results.
9. An attempt was made to examine the perceived organizational climate, leadership effectiveness, self-efficacy, trust, and job involvement among virtual workers. Few other factors such as team management, personality, conflict management, job satisfaction,

communication skills, etc. could have been included as they also affect virtual organizations in one way or the other.

10. The study could yield more consistent results if a longitudinal study would have been carried out with a limited number of virtual workers.

11. The study could yield more interesting results if factor analysis would have been carried out for results and discussion.

14.2 IMPLICATIONS

14.2.1 FOR POLICYMAKERS (ORGANIZATIONS, HR PERSONS, AND MANAGERS)

1. After this study, policymakers are able to know how to incorporate virtual working where and when as most research has been done on face-to-face teams; so for virtual workforce, rules and theories cannot be same. And yes, virtual working is not every man's cup of tea. So, policymakers need to know individual differences.

2. Virtual working offers the individual the potential advantages of flexibility, the comforts of home, and the choice of locations consistent with the individual's lifestyle. In terms of advantage to the organization, this alternative often produces cost savings and efficiency as well as employees satisfaction. Flexibility of telecommuting also allows people to deal emotional responsibilities with ease. At times, this gives individuals a daily choice in their timing of work commitment. Policymakers can promote it as an option to workers so that maximum benefit can be taken by organizations as well as workers. Policymakers can utilize this workforce for organization's maximum output by giving them virtual work option because virtual work is more like flexitime at times, and it is a way to meet the demands of caring for elderly parents or ill family members; it is even a way to better attend to such personal affairs as medical and dental appointment, home emergencies, banking needs, and so on. People are resources and assets of organization; if they are stress free, they will be definitely giving their maximum to the organizations.

3. Due to the nature of the virtual work, employees face virtual distance which affects their performance and job involvement at times. A model can be prepared using suggestion to decrease virtual distance between virtual workers which can improve affiliation need, belongingness of virtual worker to the organization, and build trust, leadership, and organizational climate in these organizations.

4. To reduce the problem of leadership crisis, new leadership styles such as authentic leadership or servant leadership can be modified to accommodate shortcomings of different leadership styles. Leaders need to play a nurturing role while being task oriented at the same time. To tackle the problem of 24 × 7 working of virtual workers, leaders can help in scheduling virtual workers' timings accordingly so that virtual work can be advantageous and satisfactory of virtual worker.

5. For developing group cohesiveness and building trust in virtual workers which will enhance virtual worker's job involvement and self-efficacy, policymakers can do the following:

    I.   encourage interviews to be taken not only on the basis of teleconferencing but face to face

    II.   plan long orientation and induction programs after recruitment

    III. organize frequent or monthly face-to-face meetings to build trust and relationship among virtual workers of organization

    IV.   give option like what IBM and other multinational companies are giving to their employees; they can come to office if they wish, so basically temporary cabins and office space can be given to virtual workers so that they can get in touch with their coworkers

    V.   encourage some more alternative work arrangements such as *Compressed Work Week*. A compressed work week is any scheduling of work that allows a full-time job to be completed in fewer than the standard 5 days. It can be useful option for workaholics who feel stressed after some time and complain of 24 × 7 working in virtual work setting organizations. This will enhance job involvement and work–life balance in virtual workers

    VI.   combine another work arrangement *Job Sharing* with virtual work to develop sense of responsibility and trust among virtual workers. In job sharing, one full-time job is assigned to two or more persons who then divide the work according to agreed-upon hours

VII. take action and efforts to sustain and develop trust that is initially developed by organizing informal meetings such as social events, celebrating birthdays, anniversary together by which team members can maintain strong social bonds. Team-building activities such as outbound training can be given to build trust and social bonds. Social and cultural events and activities can be organized by the organizations to decrease virtual distance

VIII. prepare and facilitate some recreational activities for virtual workers to develop sense of affiliation among them. One of them could be international holiday for virtual workers' families, clubs, etc.

IX. Understanding and analyzing virtual distance in context of this study is one important step that leaders can take to be more effective in managing organizations in the virtual, digital world.

KEYWORDS

- policymakers
- team-building activities
- job sharing
- team management
- conflict management

REFERENCES

Agho, A.O., Mueller, C.W., and Price, J.L. (1993). Determinants of employee job satisfaction: An empirical test of a causal model. *Human Relations*, 46(8), 1007–1027.

Akkermans, H.J.L., Isaksen, S.G., and Isaksen, E.J. (2008). Leadership for Innovation: A Global Climate Survey. A CRU Technical Report. Creativity Research Unit, the Creative Problem Solving Group, Inc., Buffalo, NY.

Akso, L. and Maryott, K.M. (2006). The relationship of employee perception of organizational climate to business-unit outcomes: An MPLS approach. *Journal of Service Research*, 18, 39–40.

Al-Alawi, A.I., Al-Marzooqi, N.Y., and Mohammed, Y.F. (2007). Organizational culture and knowledge sharing: Critical success factors. *Journal of Knowledge Management*, 11(2), 22–42. doi: 10.11.08/36732700710738818.

Allam, Z. (2002). A study of job involvement among bank employees as related to job anxiety, personality characteristics and job burnout. Unpublished doctoral thesis, Department of Psychology, Aligarh Muslim University, Aligarh.

Allam, Z. A. (2007). Study of relationship of job anxiety and job burnout with job involvement among bank employees. *Management and Labour Studies*, 21(1), 30–38.

Anderson, N.R., and West, M.A. (1998). Measuring climate for work group innovation: Development and validation of the team climate inventory. *Journal of Organizational Behaviour*, 19, 235–258.

Andriessen, J.H.E., and Vartiainen, M. (2006). Mobile Virtual Work: A New Paradigm. Heidelberg: Springer, pp. 231–252.

Aryee, S., Budhwar, P.S., and Chen, Z.X. (2002). Trust as a mediator of the relationship between organizational justice and work outcomes: Test of a social exchange model. *Journal of Organizational Behavior*, 23(3), 267–285. doi: 10.1002/job.138.

Ashok, R. (2002). Employee commitment. *Human Capital*, 5(8), 20–22.

Avolio, B.J., Kahai, S.S., and Dode, G.E. (2001). E-leadership: implications for theory, research and practice. *Leadership Quarterly*, 11, 615–668.

Avolio, B.J., Zhu, W., Koh, W., and Bhatia, P. (2004). Transformational leadership and organizational commitment: Mediating role of psychological empowerment and moderating role of structural distance. *Journal of Organizational Behavior*, 25(8), 951–968. doi: 10.1002/job.283.

Azeem, S.A. (2010). Personality hardiness, job involvement and job burnout among teachers. *International Journal of Vocational and Technical Education*, 2(3), 36–40. http://www.academicjournals.org/IJVTE.

Baba, M.L., Gluesing, J., Ratner, H., and Wagner, K.H. (2004). The contexts of knowing: Natural history of a globally distributed team. *Journal of Organizational Behavior*, 25(5), 547–587.

Bailey, M., and Luetkehans, L. (1998). Distance Learning '98. *Proceedings of the Annual Conference on Distance Teaching and Learning*, 14th, Madison, WI, August 5–7, 1998.

Bailey, D.E., and Kurland, N.B. (2002). A review of telework research: Findings, new directions, and lessons for the study of modern work. *Journal of Organizational Behavior*, 23(4), 383–400.

Balay, R. (2000). ÖrgütselBağlılık. Ankara: Nobel YayınDağıtım.

Bandura, A. (1977). Self-efficacy: Toward a unifying theory of behavioral change. *Psychological Review*, 84(2), 191–215.

Bandura, A. (1992). Exercise of personal agency through the self-efficacy mechanisms. In R. Schwarzer (Ed.), Self-efficacy: Thought Control of Action. Washington, DC: Hemisphere.

Bandura, A. (1994). Self-efficacy. In V.S. Ramachaudran (Ed.), Encyclopedia of Human Behavior, 4. New York: Academic Press, pp. 71–81.

Bandura, A. (1995). Self-efficacy in Changing Societies. New York: Cambridge University Press.

Barnatt, C. (1997). Virtual organization in the small business sector: The case of Cavendish management resources. *International Small Business Journal*, 15(4), 36–47.

Barner, R. (1996). New millennium workplace: Seven changes that will challenge managers and workers. *The Futurist*, 30(2), 14–18.

Barrett, D.J. (2006). Leadership communication: A communication approach for senior-level managers. Handbook of Business Strategy. Houston: Emerald Group Publishing, pp. 385–390.

Barrett, A. and Beeson, J. (2002). Developing Business, Leaders for 2010. The Conference Board, New York. http://www.s.op.org/tip/oct09/04konczak.aspx [Retrieved on 25 May 2012].

Bashaw, R.E., and Grant, E.S. (1994). Exploring the distinctive nature of work commitments: Their relationships with personal characteristics, job performance, and propensity to leave. *Journal of Personal Selling and Sales Management*, 14(2), 1–16.

Belanger, F., and Jordon, D. (2000). Evaluation and Implementation of Distance Learning: Technologies, tools and techniques. Hershey, PA: Idea Group Publishing.

Bell, B.S., and Kozlowski, S.W. (2002). A typology of virtual teams: Implications for effective leadership. *Group and Organization Management*, 27(1), 14–49.

Bennis, W. (1989). On Becoming a Leader. Reading, MA: Addison-Wesley Publishing Co., Inc.

Bensaou, M. (1997). Interorganizational cooperation: The role of information technology. *Information System Research*, 8(2), 107–124.

Berson, Y., Oreg, S., and Dvir, T. (2008). CEO values, organizational culture and firm outcomes. *Journal of Organizational Behavior*, 29(5), 615–633.

Birkinshaw, J. (23 Jan, 2011). Swimming lessons. Retrieved from http://bsr.london.edu/blog/post-14/index.html.

Blackwell-Shivers, S. (2004). Organizational structure and culture on leadership role requirement: The moderation impact of locus of control and self-monitoring. *Journal of Leadership and Organizational Studies*, 12(2006), 27–29.

Blau, G.J. (1986). Job involvement and organizational commitment as interactive predictors of tardiness and absenteeism. *Journal of Management*, 12(4), 577–584.

Blau, G.J., and Boal, K.B. (1987). Conceptualizing how job involvement and organizational commitment affect. *Academy of Management Review*, 12(2), 288–300.

Blau, G., and Boal, K.B. (1989). Using job involvement and organizational commitment interactively to predict turnover. *Journal of Management*, 15(1), 115–127.

Blau, G., and Ryan, J. (1997). On measuring work ethic: A neglected work commitment facet. *Journal of Vocational Behavior*, 51, 435–448.

Bleecker, S.E. (1994). Virtual organization. *The Futurist*, 28(2), 9–14.

Bommer, W.H., Rich, G.A., and Rubin, R.S. (2005). Changing attitudes about change: Longitudinal effects of transformational leader behavior on employee cynicism about organizational change. *Journal of Organizational Behavior*, 26, 733–753. doi: 10.1002/job.342.

Bowman, G.W., Jones, L.W., Peterson, R.A., Gronouski, J.A., and Mahoney, R.M. (1964). What helps or harms promotability? *Harvard Business Review*, 42(1), 6.

Brewer, A.M., and Lok, P. (1995). Managerial strategy and nursing commitment in Australian hospitals. *Journal Advanced Nursing*, 21, 769–799.

Brooke, P.P., and Price, J.L. (1989). The determinants of employee absenteeism. An empirical test of a causal model. *Journal of Occupational Psychology*, 62, 1–19.

Brooks, I., and Swailes, S. (2002). Analysis of the relationship between nurse influences over flexible working and commitment to nursing. *Journal of Advanced Nursing*, 38(2), 117–126.

Brown, S.P. (1996). A meta-analysis and review of organisational research on job involvement. *Psychological Bulletin*, 120, 235–255. http://dx.doi.org/10.1037/0033-2909.120.2.235.

Brown, S.P., and Leigh, T.W. (1996). A new look at psychological climate and its relationship to job involvement, effort and performance. *Journal of Applied Psychology*, 81, 358–368. http://dx.doi.org/10.1037/0021-9010.81.4.358.

Brown, F.W., and Moshavi, D. (2005). Transformational leadership and emotional intelligence: A potential pathway for an increased understanding of interpersonal influence. *Journal of Organizational Behavior*, 26(7), 867–871. doi: 10.1002/job.334.

Brown, S.P., and Peterson, R.A. (1993). Antecedents and consequences of salesperson job satisfaction: Meta-analysis and assessment of causal effects. *Journal of Marketing Research*, 30, 63–77.

Brunett, D., Keith, S., and Allam, E. (2002). Perceived organizatinal climate and Internet self efficacy: identifying the best climate to promote E learning, ERIC. *E-Learn 2002 World Conference on E-Learning in Corporate, Government, Healthcare, & Higher Education Proceedings* (7th, Montreal, Quebec, Canada, October 15–19, 2002). Retrieved from http://eric.ed.gov/?id=ED479438.

Bryman, A. (1992). Charisma and Leadership in Organizations. London: Sage.

Bushell, H.M. (2007). Quantifying the Key Leadership Behaviours for Creating a Successful Culture which Empowers Employees and Strengthens Organisational Performance. *Health, Work and Wellness Conference*, Toronto, Canada.

Cable, D.M., and Judge, T.A. (2003). Managers' upward influence tactic strategies: The role of manager personality and supervisor leadership style. *Journal of Organizational Behavior*, 24, 197–214. doi: 10.1002/job.183.

Cascio, W.F. (2000). Managing a virtual workplace. *The Academy of Management Executive*, 14(3), 81–90. http://www.slis.indiana.edu/faculty/hrosenba/www/l574/pdf/cascio_virtual-workplace.pdf [Retrieved on 26 Jan 2011].

Cascio, W.F., and Shurygailo, S. (2003). E leadership and virtual teams. *Organizational Dynamics*, 31(4), 362–376.

Casimir, G., Waldman, D.A., Bartram, T., and Yang, S. (2006). Trust and the relationship between leadership and follower performance: Opening the Black Box in Australia and China. *Journal of Leadership and Organizational Studies*, 12, 72–88.

Chathoth, P.K., Mak, B., Jauhari, V., and Manaktola, K. (2007). Employees perceptions of organizational trust & service climate: A structural model combining their effects on employee satisfaction. *Journal of Hospitality and Tourism Research*, 31(3), 338–357.

Chattopadhaya, S.N., and Aggarwal, K.C. (1976). Organizational Climate Inventory. Agra: National Psychological Corporation.

Cheney, P.H., and Dickson, G.W. (1982). Organizational characteristics and information systems: An exploratory investigation. *Academy of Management Journal*, 25(1), 170–184.

Choi, J.N., Price, R.H., and Vinokur, A.D. (2003). Self-efficacy changes in groups: Effects of diversity, leadership, and group climate. *Journal of Organizational Behavior*, 24(4), 357–372.

Choudhary, G. (2011). The Dynamics of Organizational Climate: An Exploration. *Management Insight*, 7(2), 111–115.

Clark, M.C., and Payne, R.L. (1997). The nature and structure of workers' trust in management. *Journal of Organizational Behavior*, 18(3), 205–224.

Cohen, A. (1999). Relationships among the five forms of commitment: An empirical analysis. *Journal of Organizational Behavior*, 20, 285–308.

Cohen, S., and Wills, T.H. (1985). Stress, social support and the buffering hypothesis. *Psychological Bulletin*, 98, 310–357.

Cohen, S., Mermistein, R., Kmatrek, T., and Hoberman, H.N. (1985). Measuring the functional components of social support. In H. Sarason and B. Sarason (Eds.) Social Support Theory, Research and Application. Dordrecht, The Netherlands: Martinus Nihoff.

Coleman, J. S. (1990). Foundations of Social Theory. Cambridge, MA: Harvard University Press.

Conger, J.A., Kanungo, R.N., and Menon, S.T. (2000). Charismatic leadership and follower effects. *Journal of Organizational Behavior*, 21, 747–767. doi: 10.1002/1099-1379(200011) 21:7 <747::AID-JOB46>3.0.CO;2-J.

Cooper, C.D., and Kurland, N.B. (2002). Telecommuting, professional isolation, and employee development in public and private organizations. *Journal of Organizational Behavior*, 23(4), 511–532.

Crant, J.M., and Bateman, T.S. (2000). Charismatic leadership viewed from above: The impact of proactive personality. *Journal of Organizational Behavior*, 21, 63–75. doi: 10.1002/(SICI)1099-1379(200002)21:1<63::AID-JOB8>3.0.CO;2-J.

Dailey, R.C., and Morgan, C.P. (1978). Personal characteristics and job involvement as antecedents of boundary spanning behaviour: A path analysis. *Journal of Management Studies*, 15(3), 330–339. doi: 10.1111/j.1467-6.

Davidson, M.C.G., and Manning, M.L. (2004). Organizational climate of food and beverage employees: Its impact upon customer satisfaction in hotels. *International Journal of Hospitality and Tourism Administration*, 4(4), 85–100.

Davis, W.D., Fedor, D.B., Parsons, C.K., and Herold, D.M. (2000). The development of self-efficacy during aviation training. *Journal of Organizational Behavior*, 21(8), 857–871.

Denison, D.R. (1996). What is the difference between organizational culture and organizational climate? A native point of view on a decade of paradigm wars. *The Academy of Management*, 21(3), 619–624.

DeSanctis, G., and Monge, P. (1999). Communication processes for virtual organizations. *Organization Science*, 10, 693–703.

Diefendorff, J., Brown, D., Kamin, A., and Lord, B., (2002). Examining the roles of job involvement and work centrality in predicting organizational citizenship behaviours and job performance. *Journal of Organizational Behaviour*, 23, 93–108.

Dienhart, J.R., and Gregoire, M.B. (1993). Job satisfaction, job involvement, job security and customer focus of quick service restaurant employee. *Journal of Hospitality and Tourism Research*, 16(2), 29–43. http://jht.sagepub. com/content/16/2/29.abstract [Retrieved on 23 June 2011].

Dimitriades, Z.S. (2007). The influence of service climate and job involvement on customer-oriented organizational citizenship behavior in Greek service organizations: A survey. *Employee Relation*, 29(5), 469–491.

Dirks, K.T., and Donald, D.L. (2002). Trust in leadership: Meta analytic findings and implications for research and practice. *Journal of Applied Psychology*, 87, 611–628.

Dirks, K.T., and Ferrin, D.L. (2001). The role of trust in organizational settings. *Journal of Organization Science*, 17(4), 65.

Douglas, C., and Gardner, W.L. (2004). Transition to self-directed work teams: Implications of transition time and self-monitoring for managers' use of influence tactics. *Journal of Organizational Behavior*, 25(1), 47–65. doi: 10.1002/job.244.

Drori, G. S., Meyer, J. W., and Hwang, H. (2006). Globalization and Organization: World Society and Organizational Change. New York: Oxford University Press.

Dubin, R. (1956). Industrial workers' worlds: A study of the central life interests of industrial workers. *Social Problems*, 3, 131–142.

Durand, S.R. (2008). Deaf Identity, Career related self efficacy and perception of career barriers of deaf and hard-of-hearing individuals. ETD Collection for Fordham University Paper AAI3312052. http://fordham.bepress.com/dissertations/AAI3312052/ [Retrieved on 4 May 2012].

Dutton, J.E., Dukerich, J.M., and Harquail, C.V. (1994). Organizational images and member identification. *Administrative Science Quarterly*, 39, 239–263.

Ebrahimi, M., and Mohamadkhani, K. (2013). The relationship between organizational climate and job involvement among teachers of high schools in Delijan City (Iran). *International Journal of Management and Research*, 4(1), 65–72.

Feghhi, J., Feghhi, J., and Williams, P. (1998). Digital certificates: Applied Internet Security. In Gerck, E. (ed.) Trust Points. Reading: Addison-Wesley, ISBN 0-201-30980-7.

Ehsan, N., Mirza, E., and Ahmed, M. (2008). Impact of computer mediated communication on virtual team's performance: An empirical study. *World Academy of Science, Engineering and Technology*, 42, 694–703.

Ekvall, G., and Ryhammar, L. (1998). Leadership styles, social climate and organizational outcomes: A study of a Swedish University College. *Leadership Style, Social Climate and Organizational Outcome*, 7(3), 126.

Ellonen, R., Blomqvist, K., and Puumalainen, K. (2008). The role of trust in organizational innovativeness. *European Journal of Innovation Management*, 11(2), 160–181.

Emery, C.R., and Barker, K.J. (2007). Effect of commitment, job involvement and teams on customer satisfaction and profit. *Team Performance Management*, 13(3/4), 90–101.

Ertürk, A. (2007). Increasing organizational citizenship behaviours of Turkish academicians: Mediating role of trust in supervisor on the relationship between organizational justice and citizenship behaviours. *Journal of Managerial Psychology*, 22(3), 257–270.

Eshraghi, H., Harati, S.H., Ebrahimi, K., and Nasiri, M. (1985). The relationship between organizational climate and leadership styles of the managers of physical education offices in Isfahan province. *Australian Journal of Basic and Applied Sciences*, 5(12), 128–130.

Euwema, M.C., Wendt, H., and van Emmerik, H. (2007). Leadership styles and group organizational citizenship behavior across cultures. *Journal of Organizational Behavior*, 28(8), 1035–1057. doi: 10.1002/job.496.

Fard, H.D., Ghatari, A.R., and Hasiri, A. (2010). Employees morale in public sector: Is organizational trust an important factor? *European Journal of Scientific Research*, 46(3), 378–390.

Feng, J., Lazar, J., and Jenny Preece, J. (2004). Empathy and online interpersonal trust: A fragile relationship. *Behaviour and Information Technology*, 23(2), 97–106. doi: 10.1080/0144929031000165924O.

Fichman, M. (2003). Straining towards trust: Some constraints essay on studying trust in organizations. *Journal of Organizational Behavior*, 24(2), 133–157. doi: 10.1002/job.189.

Fisher, K., and Fisher, M. (1998). The Distributed Mind. New York: American Management Association.

Flauto, F.J. (1999). Walking the talk: The relationship between leadership and communication competence. *Journal of Leadership Studies*, 21(3), 30–34.

Fox, L.D., Rejeski, W.J., and Gauvin, L. (2000). Effects of leadership style and group dynamics on enjoyment of physical activity. *American Journal of Health Promotion*, 14(5), 277–283.

Fukuyama, F. (1995). Trust: The Social Virtues and Creation of Prosperity. London: Hamish Hamilton.

Furnham, A., and Drakeley, R.J. (1993). Work locus of control and perceived organizational climate. *European Work and Organizational Psychologist*, 3(1), 1–9. doi: 10.1080/09602009308408572.

Furomo, K., and Pearson, J.M. (2007). Gender based communication styles. Trust and satisfaction in virtual teams. *Journal of Information, Information technology and Organization*, 2, 48–50.

Gable, M., and Dangello, F. (1994). Job involvement, Machiavellianism and job performance. *Journal of Business and Psychology*, 9(2), 159–170.

Garg, G., and Krishnan, V.R. (2003). Transformational Leadership and Organizational Structure: The Role of Value Based Leadership. http://www.rvenkat.org/garima.pdf [Retrieved on 6 May 2010].

Garman, A.N., Davis-Lenane, D., and Corrigan, P.W. (2003). Factor structure of the transformational leadership model in human service teams. *Journal of Organizational Behavior*, 24, 803–812. doi: 10.1002/job.201.

Geber, B. (1995). Virtual teams. *Training*, 32, 36–40. http://graphics.eiu.com/files/ad_pdfs/eiuForesight2020_WP.pdf.

Geisler, B. (2002). Virtual teams. 5/13/02. http://www.newfoundations.com/OrgTheory/Geisler721.html [Retrieved: 31 May 2011].

Gibb, J.R. (1991). Trust: A New Vision of Human Relationships for Business, Family and Personal Living. North Hollywood, CA: New Castle Publishing Company.

Gibson, C.B. (2001). Me and us: Differential relationships among goal-setting training, efficacy and effectiveness at the individual and team level. *Journal of Organizational Behavior*, 22(7), 789–808.

Giddens, A. (1990). The Consequences of Modernity. Oxford: Polity Press

Gilbert, S. (1996). Making the most of a slow revolution. *Change*, 28(2), 10–23.

Gist, M.E. (1987). Self-efficacy: Implications for organizational behavior and human resource management. *The Academy of Management Review*, 12(3), 472–485. Retrieved from http://www.jstor.org/stable/258514

Golden, T.D., and Raghuram, S. (2010). Teleworker knowledge sharing and the role of altered relational and technological interactions. *Journal of Organizational Behavior*, 31(8), 1061–1085. doi: 10.1002/job.652.

Goldman, J. E. (1998). Applied Data Communications, 2nd ed. New York: John Wiley and Sons, Inc.

Golden, T.D. (2006). The role of relationships in understanding telecommuter satisfaction. *Journal of Organizational Behavior*, 27(3), 319–340.

Golden, T.D., and Raghuram, S. (2010). Teleworker knowledge sharing and the role of altered relational and technological interactions. *Journal of Organizational Behavior*, 31(8), 1061–1085.

Gomez, C., and Rosen, B. (2001). The leader-member exchange as a link between managerial trust and employee empowerment. *Group and Organization Management*, 26(1), 53–69.

Great Workplace Blog. (September 1, 2009). What is Organizational Climate and Why Should You Warm Up to it? http://greatworkplace.wordpress.com/2009/09/01/what-is-organizational-climate and-why-should-you-warm-up-to-it/. Retrieved on June1, 2013.

Griffin, M.A., Hart, P.M., and Wilson-Evered, E. (2000). Using employee opinion surveys to improve organizational health. In Murphy, L.R., and Cooper, C.L. (Eds.), Health and Productive Work: An International Perspective. London: Taylor & Francis.

Griffin, M.L., Lambert, E.G., Tucker-Gail, K.A., and Baker, D.N. (2009). Job involvement, job stress, job satisfaction, and organizational commitment and the burnout of correctional staff. *Criminal Justice and Behavior,* 37(2), 239–255.

Groves, J.L., and Cho, S. (2008). Leadership styles of foodservice managers' and subordinates' perceptions. *Journal of Quality Assurance in Hospitality and Tourism*, 9(4), 317–336. doi: 10.1080/15280080802520529.

Grundy, J. (2004). Keynote presentation to GOING VIRTUAL – The Future of Work. The First Asia-Pacific Conference on Remote, Virtual Working. Brisbane, August 26–27.

Guri-Rosenblit, S. (1999). Distance and Campus Universities: Tensions and Interactions. New York: International Association of Universities and Elsevier Science Ltd.

Gurin, G., Veroff, J., and Field, S. (1960). Americans view their mental health. New York: Basic Books.

Gushue, G.V., Scanlan, K.R.L., Pantzer, K.M., and Clarke, C.P. (2006). The relationship of career decision-making self efficacy, vocational identity and career exploration behaviour in African American high school students. *Journal of Career Development*, 33(1), 19–28.

Guthrie, J.P. (2001). High-involvement work practices, turnover, and productivity: Evidence from New Zealand. *Academy of Management Journal*, 44(1), 180–190.

Hackett, R.D., Lapierre, L.M., and Hausdorf, P.A. (2001). Understanding the links between work commitment constructs. *Journal of Vocational Behavior*, 58, 392–413.

Hackman, J., and Lawler, E. (1971). Employee reactions to job characteristics. *Journal of Applied Psychology*, 52, 259–286. (Monograph).

Hall, H.H. (1987).Organizations: Structure, processes, and outcomes, 4th Ed. Englewood Cliffs, NJ: Prentice-Hall.

Hambley, L.A., O'Neill, T.A., and Kline, T.J.B. (2007). Virtual team leadership: The effects of leadership style and communication medium on team interaction styles and outcomes. *Organizational Behavior and Human Decision Processes*, 103(1), 1–20.

Handy, C. (1995). Trust and the virtual organization. *Harvard Business Review*, 73(3), 40–50.

Hannah, S.T., Woolfolk, R.L., and Lord, R.G. (2009). Leader self-structure: A framework for positive leadership. *Journal of Organizational Behavior*, 30, 269–290. doi: 10.1002/job.586.

Hao, C.C., Jung, H.C., and Yenhui, O. (2009). A study of the critical factors of the job involvement of financial service personnel after financial tsunami: Take developing market (Taiwan) for example. *African Journal of Business Management*, 3(12), 798–806.

Hare, L.R., and O'-Neil, K. (2000). Effectiveness and efficiency in small academic peer groups: A case study. *Small Group Research*, 31(1), 24–53.

Hart, P.M., Griffin, M.A., Wearing, A.J., & Cooper, C. L. (1996). Manual for the QPASS Survey. Brisbane: Public Sector Management Commission.

Hassan, A., and Ahmed, F. (2011). Authentic leadership, trust and work engagement. *International Journal of Human and Social Sciences*, 6(3), 164–170.

Hempel, P.S., Zhang, Z.-X., and Tjosvold, D. (2009). Conflict management between and within teams for trusting relationships and performance in China. *Journal of Organizational Behavior*, 30(1), 41–65. doi: 10.1002/job.540.

Henttonen, K., and Blomqvist, K. (2005). Managing distance in a global virtual team: The evolution of trust through technology-mediated relational communication. *Strategic Change*, 14, 107–119.

Hicks, J.M. (2011). Leader communication styles and organizational health. *The Health Care Manager*, 30(1), 86–91.

Hill, N.S. (2005). Leading Together, Working Together: The Role of Team Shared Leadership in Building Collaborative Capital in Virtual Teams. In Beyerlein, M.M., Beyerlein, S.T., and Kennedy, F.A. (ed.) Collaborative Capital: Creating Intangible Value (Advances in Interdisciplinary Studies of Work Teams, Volume 11). Emerald Group Publishing Limited, pp.183–209.

Hirschfeld, R.R., and Field, H.S. (2000). Work centrality and work alienation: Distinct aspects of a general commitment to work. *Journal of Organizational Behavior*, 21(7), 789–800.

Hirst, G., van Dick, R., and van Knippenberg, D. (2009). A social identity perspective on leadership and employee creativity. *Journal of Organizational Behavior*, 30(7), 963–982. doi: 10.1002/job.600.

Hmieleski, K.M., and Ensley, M.D. (2007). A contextual examination of new venture performance: Entrepreneur leadership behavior, top management team heterogeneity, and environmental dynamism. *Journal of Organizational Behavior*, 28(7), 865–889. doi: 10.1002/job.479.

Hodgetts, R.M. (1991). Organisational Behaviour: Theory and Practice. MacMillan Publishing, pp.428–430.

Howell, J.M., and Hall-Merenda, K.E. (1999). The ties that bind: The impact of leader–member exchange, transformational and transactional leadership, and distance on predicting follower performance. *Journal of Applied Psychology*, 84, 680–694.

Hsu, M.L.A., and Hsueh-Liang Fan, H.L. (2010). Organizational innovation climate and creative outcomes: Exploring the moderating effect of time pressure. *Creativity Research Journal*, 22(4), 378–386.

Huang, X., Rode, J.C., and Schroder, R.G. (2010). Organizational structures and continuous improvement and learning: Moderating effects of cultural endorsement of participative leadership. *Journal of International Business Studies*, 42(2011), 1103–1112. doi: 10.1057/jibs.2011.33.

Hunter, S.T., Bedell, K.E., and Mumford, M.D. (2007). Climate for creativity: A quantitative review. *Creativity Research Journal*, 19(1), 69–90.

Iacono, C.S., and Weisband, S. (1997). Developing trust in virtual teams. Proceedings of the Hawaii International Conference on Systems Sciences, Hawaii. (CD-ROM). http://onlinelibrary.wiley.com/doi/10.1111/j.1083-6101.1998.tb00080.x/full#b28.

Isaksen, S.G., and Ekvall, G. (2007). Assessing the Context for Change: A Technical Manual for the Situational Outlook Questionnaire. Orchard Park, NY: The Creative Problem Solving Group.

Jain, M., and Rathore, H. (2009). Technostress: Its Correlates. Unpublished Ph.D. Thesis, Rajasthan University.

Jakubowski, T.G., and Dembo, M.H. (2004). The relationship of self-efficacy, identity style and stage of change with academic staff regulation. *Journal of College Reading and Learning*, 35(1), 7–24.

Jarvenpaa, S.L., and Leidner, D.E. (1998). Communication and trust in global virtual teams. *Journal of Computer Mediated Communications*, 3(4). http://jcmc.huji.ac.il/vol3/issue4/ jarvenpaa.html [Retrieved on 21 Jan 2011].

Jaussi, K.S., Randel, A.E., and Dionne, S.D. (2007). I am, I think I can and I do: The role of personal identity, self-efficacy and cross application of experiences in creativity at work. *Creativity Research Journal*, 19(2–3), 247–258.

Jones, G.R., and George, J.M. (1998). The experience and evolution of trust: implications for co-operation and teamwork. *The Academy of Management Review*, 23(3), 531–46.

Joseph, E.E., and Winston, B.E. (2005). A correlation of servant leadership, leader trust, and organizational trust. *Leadership and Organization Development Journal*, 26(1), 6–22.

Jung, D.I., and Sosik, J. (2002). Transformational leadership in work groups: The role of empowerment, cohesiveness and collective efficacy on perceived group performance. *Small Group Research*, 33(3), 313–336. doi: 10.1177/10496402033003002.

Kahn, R.L., Wolfe, D.M., Quinn, R.P., Snoek, J.D, and Rosenthal, R.A. (1964). Organizational Stress: Studies in Role Conflict and Ambiguity. Oxford: Wiley.

Kanungo, R.N. (1979). The concepts of alienation and involvement revisited. *Psychological Bulletin*, Vol. 86, 119–38. doi: 10.1037/0033-2909.86.1.119, http://dx.doi.org/10.1037/0033-2909.86.1.119.

Kanungo, R.N. (1982). Measurement of job and work involvement. *Journal of Applied Psychology*, 67, 341–349.

Kapoor, R., and Singh, A.P. (1978). Job Involvement Scale: A pilot Study. (Unpublished). Department of Psychology, Banaras Hindu University, Varanasi.

Katzenbach, J., and Smith, D. (1993). The Wisdom of Teams. Boston: Harvard Business Press.

Kayworth, T.R., and Leidner, D.E. (2001). Leadership effectiveness in global virtual teams. *Journal of Management Information Systems*, 18(3), 7–40.

Kerlinger, F. (1973). Foundations of behavioral research. New York: Holt, Reinhart & Winston.

Kimble, C., Li, F., and Barlow, A. (2000). Effective Virtual Teams Through Communities of Practice. Unpublished manuscript, Strathclyde Business School, University of Strathclyde, Glasglow, Scotland. Available at http://www.managementscience.org/research/ab0009.asp.

Kimble, C., Feng, L., and Alexis, B. (2001) Effective Virtual Teams through Communities of Practice. Management Science Research Paper No.2000/09.

Koene, B.A.S., Vogelaar, A.L.W., and Soeters, J.L. (2002). Leadership effects on organizational climate & financial performance: Local leadership effect in chain organization. *Leadership Quarterly*, 13, 193–215.

Kramer, R.M. (1993). Cooperation and organizational identification. In Murnighan, J.K. (Ed.), Social Psychology in Organizations: Advances in Theory and Research (pp.244–268). Englewood Cliffs, NJ: Prentice-Hall.

Kunze, F., Boehm, S.A., and Bruch, H. (2011). Age diversity, age discrimination climate and performance consequences—a cross organizational study. *Journal of Organizational Behavior*, 32, 264–290. doi: 10.1002/job.698.

Lambert, E., Hogan, N.L., Barton, S.M., and Oko, E. (2009). The impact of job stress, job involvement, job satisfaction and organizational commitment on correctional staff support for rehabilitation and punishment. *Criminal Justice Studies*, 22(2), 109–122.

Lamsa, A.M., and Pucetaite, R. (2006). Developing organizational trust among employees from a contextual perspective. *Journal of Business Ethics: A European Review*, 15(2), 130–141.

Lee, Y.D., and Lin, K.T. (1999). A research on the relationships among superior's leadership style, employee's communication satisfaction and leadership effectiveness: A case study of the Taiwan sugar corporation. *Chinese Management Review*, 2, 1–19.

Lester, S., and Brower, H. (2003). In the eyes of the beholder: The relationship between subordinates' felt trustworthiness and their work attitudes and behavior. *Journal of Leadership and Organizational Studies*, 10(2), 168–179.

Lewicki, R.J., McAllister, D.J., and Bies, R.J. (1998).Trust and distrust: New relationships and realities. *The Academy of Management Review*, 23(3), 428–459.

Likert, R. (1967). The Human Organization: Its Management and Value. New York, NY: McGraw-Hill.

Lipnack, J., and Stamps. J. (1997). Virtual Teams: Reaching Across Space, Time, and Organizations with Technology. New York, NY: John Wiley and Sons, Inc.

Lipnack J., and Stamps, J. (1999). Virtual teams: The new way to go. *Strategy and Leadership*, 27 (1), 14–19. (Jan/Feb, 14–19).

Litwin, G.H., and Stringer, R.A. (1968). Motivation and Organizational Climate. USA: Harvard University Press.

Locke, E.A., and Latham, G.P. (1990). A theory of goal setting and task performance. Englewood, NJ: Prentice Hall.

Lodahl, T.M., and Kejner, M. (1965). The definitional and measurement of job involvement. *Journal of Applied Psychology*, 49, 24–33.

Lorence, J., and Mortimer, J. (1985). Job involvement through the life course: A panel study of 3 age groups. *American Sociological Review*, 50 (5), 618–638.

Lozeski, K.S. (2009). Leading the Virtual Workforce: How Great Leaders Transform Organizations in 21st Century. NJ: John Wiley & Sons.

Lussier, R.N., and Achua, C.F. (2011). Effective Leadership. New Delhi: South Western Cengage Learning.

Luthans, F. (2003). Organizational Behavior, 9th Edition. New York: McGraw-Hill, pp. 460–482, 560–572.

Luthans, F., Norman, S.M., Avolio, B.J., and Avey, J.B. (2008). The mediating role of psychological capital in the supportive organizational climate – employee performance relationship. *Journal of Organizational Behavior*, 29(2), 219–238.

Mael, F., and Ashforth, B. E. (1995). Loyal from day one: Biodata, organizational identification, and turnover among newcomers. *Personnel Psychology*, 48, 309–333.

Markman, G.D., Baron, R.A., and Balkin, D.B. (2005). Are perseverance and self-efficacy costless? Assessing entrepreneurs' regretful thinking. *Journal of Organizational Behavior*, 26(1), 1–19.

Marshall, G.W., Lassk, F.G., and Moncrief, W.C. (2004). Salesperson job involvement: Do demographic, job situational, and market variables matter? *Journal of Business and Industrial Marketing*, 19(5), 337–343.

Mashayekhi, M., Sajjadi, S.A.N., and Tabrizi, K.G. (2013). The relationship between organizational climate school & job involvement of physical education teachers. *Switzerland Research Park Journal*, 102(10), 962–963.

Mathebula, M.R.L. (2004). Modeling the Relationship Between Organizational Commitment, Leadership Style, Human Resource Practices and Organizational Trust. Ph.D dissertation, Faculty of Economics and Management Sciences, University of Pretoria, Pretoria.

Mathisen, G.E., and Einarsen, S. (2004). A review of instruments assessing creative and innovative environments within organizations. *Creativity Research Journal*, 16(1), 119–140.

Mayer, R.C., Davis, J.H., and Schoorman, F.D. (1995). An integrative model of organizational trust. *Academy of Management Review*, 20, 709–734.

McAlearney, A.S. (2006). Leadership development in healthcare: A qualitative study. *Journal of Organizational Behavior*, 27(7), 967–982. doi: 10.1002/job.417.

McDonough, E., Kahn, K., and Barczak, G. (2001). An investigation of the use of global, virtual, and collocated new product development teams. *The Journal of Product Innovation Management*, 18, 110–120.

McElroy, J.C., Morrow, P.C., Crum, M.R., and Dooley, F.J. (1995). Railroad employee commitment and work-related attitudes and perceptions. *Transportation Journal*, 13–24.

McElroy, J.C., Morrow, P.C., and Wardlow, T.R. (1999). A career stage analysis of police officer work commitment. *Journal of Criminal Justice*, 27(6), 507–516.

McShane, S.L., and Glinow, M.A.V. (2003). Organizational Behaviour: Emerging Realities for the Workplace Revolution, 2nd Edition. New Delhi: Pearson Publishers, 11, 437.

Meyerson, D., Weick, K.E., and Kramer, R.M. (1996). Swift trust and temporary groups. In Kramer, R.M., and Tyler, T.R. (Eds.) Trust in Organizations: Frontiers of Theory and Research (pp.166–195). Thousand Oaks, CA: Sage Publications.

Michaelides, M. (2008). Emerging themes from early research on self-efficacy beliefs in school mathematics. *Electronic Journal of Research in Educational Psychology*, 6(1), 219–234.

Milligan, P.K. (2003). The impact of trust in leadership on officer commitment and intention to leave military service in the U.S. air force. USA: Capella University Press.

Mishra, A.K. (1996). Organizational response to crisis: The centrality of trust. In Kramer, R.M., and Tyler, T.R. (Eds.) Trust in Organizations: Frontiers of Theory and Research, (pp.261–287). Thousand Oaks, CA: Sage Publications, Inc.

Mishra, P.C., and Shyam, M. (2005). Social support & job involvement in prison officers. *Journal of the Indian Academy of Applied Psychology*, 31(1–2), 7–11.

Mishra, A.K., and Wagh, A. (2004). A comparative study of job involvement among business executives. *Indian Journal of Training and Development*, 34(2), 79–84.

Mishra, K.E., Spreitzer, G.M., and Mishra, A.K. (1998). Preserving employee morale during downsizing. *Sloan Management Review*, (Winter), 83–95.

Misiolek, N.I., and Heckman, R. (2005). Patterns of Emergent Leadership in Virtual Teams, System Sciences, HICSS'05, *Proceeding of the 38th Annual Hawaii International Conference.*

Misztal, B.A. (1996). Trust in Modern Societies. Oxford: Polity Press

Mohammed, S., Mathieu, J.E., and Bartlett, A.L. (2002). Technical-administrative task performance, leadership task performance, and contextual performance: Considering the influence of team- and task-related composition variables. *Journal of Organizational Behavior*, 23(7), 795–814.

Morgan, G. (1997). Images of Organizations. Thousand Oaks, CA: Sage Publications.

Morris, S.A., Marshall, T.E., and Rainer, R.K. (2003). Trust and Technology in Virtual Teams in Advanced Topics in Information Resource Management. Hershey, PA, USA: IGI Publishing, pp.133–159.

Nelson and Quick. (2009). Self-Efficacy, Organizational Behaviour. New Delhi: Cengage Learning Publishers, p. 90.

Nyhan, R.C., and Marlowe, H.A. (1997). Development and psychometric properties of the organizational trust inventory. *Evaluation Review*, 21, 614–635.

Obenchain, A. (2002). Organizational Culture and Organizational Innovation in Not-for-Profit, Private and Public Institutions of Higher Education. UMI: Unpublished Dissertation, Nova Southeastern University.

O'Driscoll, M.P., and Schubert, T. (1988). Organizational climate and burnout in a New Zealand social service agency. *Work and Stress*, 2(3), 199–204. http://dx.doi.org/10.1080/02678378808259167.

Örs, M., Acuner, A.M., Sarp, N., and Önder, Ö.R. (2003). AntalyaTıpFakültesiHastanesi'nde, Antalya SosyalSigortalarKurumuHastanesi'ndeve Antalya DevletHastanesi'ndeçalışa nhekimlerilehemşirelerinörgütlerinebağlılıklarınailişkingörüşlerinindeğerlendirilmesi. *Ankara Üniversitesi Tıp FakültesiMecmuası*, 56(4), 217–224.

Ouyang, Y. (2009). The mediating effects of job stress and job involvement under job instability: Banking service personnels of Taiwan as an example. *Journal of Money, Investment and Banking*, 11, 16–27. Retrieved on 21 Jan 2016. http://www.eurojournals.com/JMIB.htm.

Özsoy, S.A., Ergül, Ş., and Bayık, A. (2004). Biryüksekokulçalışanlarınınkurumabağlılı kdurumlarınınincelenmesi.İş, Güç EndüstriİlişkileriveİnsanKaynaklarıDergisi, 6(2): 56–75.

Paglis, L.L., and Green, S.G. (2002). Leadership self-efficacy and managers' motivation for leading change. *Journal of Organizational Behavior*, 23, 215–235. doi: 10.1002/job.137.

Pathak, H. (2011). Organizational Climate, Organization Change. Delhi: Pearson Publishers, pp. 39–40.

Paul, D.L., and Mcdaniel, R.R. (2004). A field study of the effect of interpersonal trust on virtual collaborative relationship performance. *MIS Quarterly*, 28(2), 183–227.

Paullay, I., Alliger, G., and Stone-Romero, E. (1994). Construct validation of two instruments designed to measure job involvement and work centrality. *Journal of Applied Psychology*, 79, 224–228.

Peters, T. (1993). Peters on excellence. *Washington Business Journal*, 12(11), 51.

Pettigrew, A.M. (1979). On studying organizational cultures. *Administrative Science Quarterly*, 2, 570–581.

Pieterse, A.N., van Knippenberg, D., Schippers, M., and Stam, D. (2010).Transformational and transactional leadership and innovative behavior: The moderating role of psychological empowerment. *Journal of Organizational Behavior*, 31(4), 609–623. doi: 10.1002/job.650.

Pitcher, P. (1994 French). Artists, Craftsmen, and Technocrats: The Dreams Realities and Illusions of Leadership, 2nd English edition. Stoddart Publishing, Toronto, 1997. ISBN 0-7737-5854-2.

Politis, J.D. (2003). The connection between trust and knowledge management: What are its implications for team performance. *Journal of knowledge management*, 7(5), 55–66.

Ponnu, C.H., and Tennakoon, G. (2009). The association between ethical leadership and employee outcomes—The Malaysian case. *Electronic Journal of Business Ethics and Organizational Studies,* 14(1), 21–32.

Potosky, D., and Ramakrishna, H. (2001). Goal orientation, self-efficacy, organizational climate, and job performance. *Academy of Management*.

Powell, D., Piccoli, G., and Ives, B. (2004). Virtual teams: A review of current literature and directions for future research. *The DATA BASE for Advances in Information Systems*, 35(1), 6–36. ACM Press.

Prussia, G.E., Anderson, J.S., and Manz, C.C. (1998). Self-leadership and performance outcomes: The mediating influence of self-efficacy. *Journal of Organizational Behavior*, 19(5), 523–538.

Putti, J., and Song Kheun, L. (1986). Organizational climate – job satisfaction relationship in a public sector organization. *International Journal of Public Administration*, 8(3), 337–344.

Quinn, R.E. (1988). Beyond Rational Management. San Francisco, CA: Jossey-Bass.

Rabinowitz, S., and Hall, D.T. (1977). Organisational research on job involvement. *Psychological Bulletin*, 84, 265–288. http://dx.doi.org/10.1037/0033-2909.84.2.265.

Rafferty, A.E., and Rose, D.M. (2001). An examination of the relationship among extent of workplace change, employee participation, and workplace distress. *Australian Journal of Psychology*, 53, 85.

Raghuram, S., and Weisenfeld, B. (2004). Work-non work conflict and job stress among virtual workers. *Human Resource Management: The Intersection of Information Technology and HRM*, 43(2–3), 259–277. doi: 10.1002/hrm.20019.

Raghuram, S., Garud, R., Weisenfeld, B., and Gupta, V. (2001). Factors contributing to virtual work, adjustment. *Journal of Management*, 27, 383–405.

Ramsey, R., Lassk, F.G., and Marshall, G.W. (1995). A critical evaluation of a measure of job involvement: The use of the Lodahl and Kejner (1965) scale with salespeople. *Journal of Personal Selling and Sales Management*, 15 (3), 65–74.

Ratnasingam, P. (2005). E-Commerce relationship: The impact of trust on relationship continuity. *International Journal of Commerce and Management*, 15(1), 1–16.

Reichers, A.E., and Schneider, B. (1990). Climate and culture: An evolution of constructs. In Schneider, B. (Ed.) Organizational Climate and Culture (pp.5–39). San Francisco: Jossey-Bass.

Richardson, H.A., and Vandenberg, R.J. (2005). Integrating managerial perceptions and transformational leadership into a work-unit level model of employee involvement. *Journal of Organizational Behavior*, 26(5), 561–589. doi: 10.1002/job.329.

Rose, D.M., and Griffin, M. (2002). High Performance Work Systems, HR Practices and High Involvement: A group level analysis. Academy of Management, Conference 2002, Denver, USA.

Rose, D.M., and Waterhouse, J.M. (2004). Experiencing New Public Management: Employee Reaction to Flexible Work Practices and Performance Management. Industrial Relations European Conference, Utrecht, Netherlands

Rose, D.M., Douglas, M., Griffin, M.A., and Linsley, C. (2002). Making HR Work: Symposium – Managing the Relationship: Commitment and Work Effectiveness. Australian Human Resources Institute HR Practices Day 2002. Brisbane, Australia.

Rotenberry, P.F., and Moberg, P.J. (2007). Assessing the impact of job involvement on performance. *Management Research News*, 30, 203–215. http://dx.doi.org/10.1108/01409170710733278.

Rousseau, D.M., Sitkin, B.B., Burt, R.S., and Camerer, C. (1998). Not so different after all: A cross disciplinary view of trust. *The Academy of Management Review*, 23(3), 393–404.

Sagie, A., Zaidman, N., Amichai-Hamburger, Y., Te'eni, D., and Schwartz, D.G. (2002). An empirical assessment of the loose–tight leadership model: Quantitative and qualitative analyses. *Journal of Organizational Behavior*, 23, 303–320. doi: 10.1002/job.153.

Sarker, S., Lau, F., and Sahay, S. (2000). Using an adapted grounded theory approach for inductive theory building about virtual team development. SIGMIS Database vol. 32, Issue 1, p. 80.

Saskin, M. (1990). The Visionary Leader: The Leader Behavior Questionnaire. Organizational Design and Development Inc.

Schermerhorn, J.R., Hunt, J.G., and Osborn, R.N. (2010). Telecommuting. Organizational Behavior, 10th Edition. New Delhi: John Wiley & Sons, p.142.

Schultz, D., and Schultz, S.E. (2004). Psychology and Work Today: An Introduction to Industrial and Organizational Psychology, 8th Edition. Delhi: Pearson Publishers, pp. 13–14, 251.

Schwarzer, R., and Borm, A. (1997). Optimistic self-beliefs: Assessment of general perceived self-efficacy in thirteen cultures, *World Psychology*, 3(1–2), 177–190.

Sekaran, U. (1989). Paths to the job satisfaction of banking employees. *Journal of Organisational Behaviour*, 10, 347–359. http://dx.doi.org/10.1002/job.4030100405.

Sekeran, U., and Mowday, R.T. (1981). A cross cultural analysis of the influence of individual and job characteristics on job involvement. *International Review of Applied Psychology*, 30, 51–64. http://dx.doi.org/10.1111/j.1464-0597.1981.tb00979.x.

Serva, M.A., Fuller, M.A., and Mayer, R.C. (2005). The reciprocal nature of trust: A longitudinal study of interacting teams. *Journal of Organizational Behavior*, 26(6), 625–648. doi: 10.1002/job.331.

Shapiro, D.L., Furst, S.A., Spreitzer, G.M., and Von Glinow, M.A. (2002). Transnational teams in the electronic age: Are team identity and high performance at risk? *Journal of Organizational Behavior*, 23, (4), 455–467.

Shivers-Blackwell, S. (2004). Using role theory to examine determinants of transformational and transactional leader behavior. *Journal of Leadership and Organizational Studies* (Baker College), 10(3), 41–50.

Shivers-Blackwell, S. (2006). The influence of perceptions of organizational structure & culture on leadership role requirements: The moderating impact of locus of control & self-monitoring. *Journal of Leadership and Organizational Studies*, (Summer 12), 27–49, doi: 10.1177/107179190601200403.

Simmons, B.L., Gooty, J., Nelson, D.L., and Little, L.M. (2009). Secure attachment: Implications for hope, trust, burnout, and performance. *Journal of Organizational Behavior*, 30(2), 233–247. doi: 10.1002/job.585.

Singh, A.P. (1984). Job Involvement Scale. MannualAbhishek Publishers.

Singh, K. (2010). Leadership, Organizational Behavior Text and Cases. Delhi: Pearson Publishers, pp.260–271.

Sjöberg, A., and Sverke, M. (2000). The interactive effect of job involvement and organizational commitment on job turnover revisited: A note on the mediating role of turnover intention. *Scandinavian Journal of Psychology*, 41, 247–252.

Skaalvik, E.M., and Skaalvik, S. (2009).Teacher self-efficacy and teacher burnout: A study of relations. *Teaching and Teacher Education*, 26, 1059–1069.

Smith, G. (2005). How to achieve organizational trust within an accounting department. *Managerial Auditing Journal*, 20(5), 520–532.

Somech, A. (2003). Relationships of participative leadership with relational demography variables: A multi-level perspective. *Journal of Organizational Behavior*, 24, 1003–1018. doi: 10.1002/job.225.

Sosik, J.J., and Godshalk, V.M. (2000). Leadership styles, mentoring functions received, and job-related stress: a conceptual model and preliminary study. *Journal of Organizational Behavior*, 21, 365–390. doi: 10.1002/(SICI)1099-1379(200006)21:4<365::AID-JOB14>3.0.CO;2-H.

Sosik, J.J., Avolio, B.J., and Kahai, S.S. (1997). Effects of Leadership style and anonymity on group potency and effectiveness in a group decision support system environment, *Journal of Applied Psychology*, 82(1), 89–103.

Sparks, J.R., and Schenk, J.A. (2001). Explaining the effects of transformational leadership: An investigation of the effects of higher-order motives in multilevel marketing organizations. *Journal of Organizational Behavior*, 22(8), 849–869. doi: 10.1002/job.116.

Spreitzer, G. (2007). Giving peace a chance: Organizational leadership, empowerment, and peace. *Journal of Organizational Behavior*, 28(8), 1077–1095.

Spreitzer, G.M., Perttula, K.H., and Xin, K. (2005). Traditionality matters: An examination of the effectiveness of transformational leadership in the United States and Taiwan. *Journal of Organizational Behavior*, 26, 205–-227.

Sproull, L., and Kiesler, S. (1991). Connections: New Ways of Working in the Networked Organization. Cambridge, MA: MIT Press.

Srivastava, S.K. (2001). Job involvement and mental health among executive and supervisors. *Journal of Community Guidance*, 18(3), 365–372.

Stam, D.A., van Knippenberg, D., and Wisse, B. (2010). The role of regulatory fit in visionary leadership. *Journal of Organizational Behavior*, 31(4), 499–518. doi: 10.1002/job.624.

Staples, D.S., Hulland, J.S., and Higgins, C.A. (1998). A self-efficacy theory explanation for the management of remote workers in virtual organizations. *Journal of Computer Mediated Communication*, 3(4), 29. doi: 10.1111/j.1083-6101.1998.tb00085.x.

Steward, R.B. (2004). Employee Perceptions of Trust: Rebuilding the Employee-Employer Relationship. Doctor of Philosophy, Regent University, England

Steward, R. (2006). Virtual Teams in Team Management, 1st Edition, New Delhi: Infinity Books, pp.462–470.

Strachan, S.M, Woodgate, J., Brawley, L.R., and Tse, A. (2005). The relationship of self-efficacy and self-identity to long-term maintenance of vigorous physical activity. *Journal of Applied Biobehavioral Research*, 10(2), 98–112.

Suchan, J., and Hayzak, G. (2001). The communication characteristics of virtual teams: A case study. *IEEE Transactions on Professional Communication*, 44 (3), 174.

Sullivan J., Peterson, R.B., Kameda, N., and Shimada, J. (1981). The Relationship Between Conflict Resolution Approaches and Trust-A Cross Cultural Study. *Academy of Management Journal*, 24(4), 803–815.

Tagiuri, R. (1968). The concept of organizational climate. In Tagiuri, R., and Litwin, G.H. (Eds.) Organizational Climate: Exploration of a Concept. Boston: Harvard University, Division of Research, Graduate School of Business Administration.

Taj, H. (2001). Leadership Effectiveness Scale. Agra: National Psychological Corporation.

Tata, J., Prasad, S., and Thorn, R. (1991). The influence of organizational structure on the effectiveness of TQM programs, *Journal of Managerial*. http://www.questia.com/googlescholar.qst [Retrieved on 7 April 2012].

Thomas, J.C. (2008). Administrative, faculty and staff perceptions of organizational climate and commitment. *Christian Higher Education*, 7(3).

Thoms, P., Moore, K.S., and Scott, K.S. (1996). The relationship between self-efficacy for participating in self-managed work groups and the big five personality dimensions. *Journal of Organizational Behavior*, 17(4), 349–362.

Townsend, A.M., DeMarie, S., and Hendrickson, A.R. (1998). Virtual teams: Technology and the workplace of the future. *Academy of Management Executive*, 12(3), 17–29.

Uygur, A., and Gonca, G. (2009). A study into organizational commitment and job involvement: An application towards the personnel in the central organization for Ministry of Health in Turkey. *Ozean Journal of Applied Sciences*, 2(1), 113–125.

van Breukelen, W., van der Leeden, R., Wesselius, W., and Hoes, M. (2010). Differential treatment within sports teams, leader–member (coach–player) exchange quality, team atmosphere, and team performance. *Journal of Organizational Behavior*, 33(1), 43–63. doi: 10.1002/job.735.

Van Dusen, Gerald C. (1997). The Virtual Campus: Technology and Reform in Higher Education. ASHE-ERIC Higher Education Report, Volume 25, No 5. Washington DC.

Van Dyne, L., Vandewalle, D., Kostova, T., Latham, M.E., and Cummings, L.L. (2000). Collectivism, propensity to trust and self-esteem as predictors of organizational citizenship in a non-work setting. *Journal of Organizational Behavior,* 21(1), 3–23.

Vlaar, P.W.L., VandenBosch, F.A.J., and Volberda, H.W. (2007). On the evolution of trust, distrust and formal coordination and control in interorganizational relationships: Toward an integrative framework. *Group Organization Management*, 32(4), 407–428. doi: 10.1177/1059601106294215.

Vroom, V.H. (1962). Ego-involvement, job satisfaction, and performance. *Personal Psychology*, 15, 159–177.

Vries, R. Ede, Baker-Piper, A., and Oostenveld (2009). Leader communication? The relations of leader's communication styles with leadership styles, knowledge sharing and leadership outcomes. *Journal of Business Psychology*, 25(3), 67–38.

Walumbwa, F.O., Luthans, F., Avey, J.B., and Oke, A. (2009). Authentically leading groups: The mediating role of collective psychological capital and trust. *Journal of Organizational Behavior*, 32(1), 4–24. doi: 10.1002/job.653. http://cba.unl.edu/research/articles/1352/download.pdf [Retrieved on 23 June 2009].

Wang, Y. (2003). Trust and decision-making styles in Chinese township-village enterprises. *Journal of Managerial Psychology*, 18(6), 541–556.

Warkentin, M., and Beranek, P.M. (1999). Training to improve virtual team communication. *Information Systems Journal*, 9(4), 271–289. doi: 10.1046/j.1365-2575.1999.00065.x.

Warkentin, M. E., Sayeed, L., and Hightower, R. (1997). Virtual Teams versus Face-to-Face Teams: An Exploratory Study of a Web-based Conference System. *Decision Sciences*, 28: 975–996.doi: 10.1111/j.1540-5915.1997.tb01338.x.

Watson-Manheim, M. B., and Belanger, F. (2002). Support for communication-based work processes in virtual work. *e-Service Journal*, 61-82.

Webber, S. S., and Klimoski, R. J. (2004). Client–project manager engagements, trust, and loyalty. *Journal of Organizational Behavior*, 25(8), 997–1013. doi: 10.1002/job.294.

Wech, B. (2002). Trust context: Effect on organizational citizenship behavior, supervisory fairness and job satisfaction beyond the influence of leader-member exchange. *Business and Society*, 41(3), 353–360.

Wellman, B. (2001). The rise of networked individualism. In Keeble, L. (Ed.) Community Networks Online (pp.17–42). London: Taylor & Francis.

Wilson, M.B. (1993). A New Method of Assessing Cook and Wall's Informal Theory of Organizational Trust: A Coast Guard Sample. (Doctoral Dissertation, George Washington University,1993). Dissertation Abstract International, 54, 109.

Witt, L. (1993). Reaction to work assignments as predictors of organizational commitment: the moderating effect of occupational identification. *Journal of Business Research*, 26, 17–30.

Word, J., and Park, S.M. (2009). Working across the divide: ***Job involvement*** in the public and nonprofit sectors. *Review of Public Personnel Administration*, 29(2), 103–133.

Yilmaz, A., and Atalay, C.G. (2009). A theoretical analyze on the concept of trust in organizational life. *European Journal of Social Science*, 8(2), 14.

Yoong, P.J. (2001). Relationship building and the use of ICT in boundary-crossing virtual teams: A facilitator's perspective. *Journal of Information Technology*, 16(4), 205.

Young, J.R. (2002). 'Hybrid' teaching seeks to end the divide between traditional and online instruction. The Chronicle of Higher Education. http//.www.chronicle.com/free/ v48 /i28/28a03301.htm [Retrieved on 3 February 2011].

Yukl, G.A. (1994). Leadership in organizations. Englewood Cliffs, NJ: Prentice-Hall, New Haven, CT: Yale University Press.

Zammuto, R.F., and Krakower, J.Y. (1991). Quantitative and qualitative studies of organizational culture. *Research in Organizational Change and Development*, 5, 83–114.

Zammuto, R.F., and O'Connor, E.J. (1992). Gaining advanced manufacturing technologies' benefits: The roles of organizational design and culture. *Academy of Management Review* 17, 701–728.

Ziurs, I. (2003). Leadership in virtual teams: Oxymoron or opportunity? *Organizational Dynamics*, 31(4), 339–351.

"If You Are Going to Downsize, Says U.S. Labor Secretary Robert Reich, Do It Gently," interview, Sales & Marketing Management, volume 148, September 1996, pp. 118-123.

OTHER READINGS:

http://www.reliableplant.com/Read/12675/communication-most-imprtant-key-to-leadership-success

http://upted.up.ac.za//thesis/available/etd-0706 2004-112817/unrestricted/ oothsis.pdf

http://www.eurojournals. com/ ejsr.htm.

http://www.eurojournals.com/ejss-8-2-14.pdf

http://en.wikipedia.org/wiki/Organization_climate#See_also http:/www.Thework91.com/ articles/comchan.htm

http://www.inflibnet.ac.in

http://gom.sagepub.com/cgi/content/abstract/27/1/14

http://www.jstor.org/stable/4093782

http://www.battlebook.org/military/classes/Mentoring/Organizational_climate.pdf

http://www.qccigw.ca/uploads/userfiles/creatingpositieclimate-us-aug09.pdf

http://www.cpsb.com/research/articles/featured-articles/globalClimateSurvey-Technical-Report.pdf http://www.insipub.com/ajbas/2011/Dec 2011/ 1985-1990.pdf

http://en.wikipedia.org/wiki/trust_(sociology)

www.3interscience.wiley.com/journal/114229532/abstract

http://cmr.ba.ouhk.edu.hk/cmr/oldweb/n5/981032.html

http://www.regent.edu/acad/global/publication/sl-proceedings/2006/danhauser-boshoff.pdg

www.3interscience.wiley.com/journal/10436418/abstract

www.3interscience.wiley.com/journal/65781/abstract

http://jiito.org/articles/jiitov2po47-060furumo35.pdf

http://is2.Ise.ac.uk/asp/aspecis/20010078.pdf

www.ariadne.ac.uk/issue43/panteli/

http://www.3interscience.wiley.com/journal

http://portal.acm.org/

http://portal.acm.org/citation.cfm?id=962749.962825qandcoll=&dl=acm

www.apa.org/

http://en.wikipedia.org/wiki/virtual_team

http://www.seanet.com/~daveg/vrteams.htm

http://www.merlien.org/oj/index.php/JOE/article/view File/21/15

http://en.wikipedia.org/wiki/Organizational_climate

http://www.qualityvalues.com/organizational_audits/organizational_climate.htm

http://cnx.org/content/m13465/latest/

http://www.lib.umd.edu/groups/learning/reports/2004ocdasurvey.pdf

http://www.springelink.com/content/pk7k1160x7038350/

http://web.njit.edu/~jerry/Virtual_Team_Leadership.html

http://ieeexplore.iece.org/Xplore/login.jsp?url=/ie15/9518/30166/01385331.pdf?temp=x

http://www.leadingvirtually.com/

http://howe.stevens.edu/fileadmin/Files/News_Events/howe_forum/Leadership_in_
    Virtual_Teams.pdf

http://www.des.emory.edu/mfp/self-efficacy.html.

http://sgr.sagepub.com/cgi/content/abstract/37/1/65

http://portal.acm.org/citation.cfm?id=1277648

http://orgsci.journal.informs.org/cgi/content/abstract/10/6/758

http://www.startwright.com/virtual.htm

http://www.springerlink.com/content/yp776821v01330g3/

http://www.ifets.info/journals/8_4/19.pdf

http://medind.nic.in/jak/t05/i1/jakt05i1p7.pdf

http://www.chrs.rutgers.edu/pub_documents/Huselid_18.pdf

http://findarticles.com/p/articles/mi_qa3629/is_199506/ai_n8719596

http://en.wikipedia.org/wiki/Virtual_Team

http://prtal.acm.org

www.academicjournal.org/IJVTE

http://www.inflibnet.ac.in

INDEX